Genetics of Bacteria

Sheela Srivastava

Genetics of Bacteria

Springer

Sheela Srivastava
Department of Genetics
University of Delhi, South Campus
New Delhi, Delhi
India

ISBN 978-81-322-1089-4 ISBN 978-81-322-1090-0 (eBook)
DOI 10.1007/978-81-322-1090-0
Springer New Delhi Heidelberg New York Dordrecht London

Library of Congress Control Number: 2013930625

Printed on acid-free paper

Springer is part of Springer Science+Business Media (www.springer.com)

Preface

From the time microorganisms could be seen, described, and studied, they have provided a useful system to gain insight into the basic principles of life, though we are still far from understanding them fully. The relative simplicity, which may often be deceptive, made microbes ideally suited for answering some very fundamental questions in science. Microorganisms have been employed in almost all fields of biological studies, including the science of Genetics.

The whole edifice of classical genetics centers around three processes viz, the generation, expression, and transmission of biological variation. Thus, the most crucial requirement of genetic analysis is to select or introduce variations in a specific gene (mutation\s). Even with the rapid growth of modern molecular biology, the relevance of genetic analyses that depends on finding the mutants and using them to elucidate the normal structure and operation of a biological system has not been lost. Mutation may lead to an altered trait in an organism and if the change takes place in the observable characteristics (phenotype) of that organism, this could be used to follow the transmission of the said gene. The genetic composition (genotype) of the organism could be inferred from the observable characteristics. With the improved biochemical techniques and instrumentation, and the revelation of the chemical nature of the gene (DNA, RNA in some viruses), it became possible to dissect the gene expression/function at the molecular level. In all these studies, gene mutations occupied the center stage.

Gene cloning and other techniques of gene manipulation provided a new direction, as the genes could be isolated and studied without a prior requirement of obtaining mutants. Moreover, these genes could be altered specifically and in a desired manner in vitro (site-directed mutagenesis). However, deriving the gene function by its alteration never lost its relevance. Also, it should become clear that most genes do not function in isolation and its real understanding can come through in vivo analysis only.

When microorganisms, first fungi and then bacteria, were employed as model systems in genetic analysis, they offered immediate advantages in studying all the three aspects of heredity: being haploid and structurally simpler it became easy to isolate mutations of various kinds and relate them to a specific function. Though very few morphological mutants could be obtained, a whole range of biochemical mutants became available in a very short time. The availability of these mutants and their amenability to detail genetic and biochemical analyses led to the generation of a whole lot of information about gene expression and its regulation. So much so, that they provided the first clues, and the platform for studying the complex eukaryotic systems.

It was when transmission of biological variation was to be studied that a different strategy had to be employed. While in higher organisms, such a line of study would require phenotypic markers in a controlled hybridization, in microorganisms, especially bacteria, a more genetic approach needed to be employed.

Both bacteria and their viruses and fungi have been extensively exploited for genetic analyses. The information so generated became so vast that creation of a branch of Microbial Genetics became thoroughly justified. Microorganisms have not attempted to alter any established genetic concept but the technique applied to them and the way the results are to be interpreted are so different from higher organisms that their clubbing together may cause some confusion. In the same vein, fungi and bacteria represent two entirely different types of biological systems, i.e., eukaryotic and prokaryotic, respectively. Thus, it would not be inappropriate to treat them separately.

Bacteria, the simplest of the living organisms, have provided enough material on all aspects of genetics. In any compendium, however, the treatment of these aspects may be very different. In most basic genetics books, bacterial genetics may occupy the place of a chapter with information about mutation and expression combined with other eukaryotes. Some books dealing with microbes or more correctly with bacteria alone are also available and have served the purpose of a useful resource book on Microbial Genetics to teachers as well as students. While teaching a course on Microbial Genetics for the last 25 years to post-graduate students at Delhi University, I have realized that a book on Bacterial Genetics may be very handy to students, researchers, and teachers alike. However, a new format has been planned for this book where emphasis has been on the transmission aspects, along with giving due share to the generation of biological variation, because without the latter, the former is not possible. The omission of expression part has indeed been intentional. And the reason: a large volume of information available on this aspect in books dealing with genetics, biochemistry, cells biology, molecular biology, and biotechnology. Thus, the inclusion of such information would only have amounted to repetition.

Bacterial genetics is moving through an important phase in its history. While on the one hand, this field of study continues to remain instrumental in the development of new tools and methodologies for better understanding of molecular biology, on the other, it provides scientists with a strong handle whose ultimate impact is hard to foresee. In addition to providing an insight into basic biological questions, genetic knowledge can also be used to manipulate biological systems for scientific or economic reasons. Traditionally, genetic manipulation requires mutagenesis, gene transfer, and genetic recombination followed by selection for desired characteristics. However, with such techniques, geneticists are forced to work with random events with selection often quite complex to detect rare mutations with the desired genotypes. Moreover, the nature of the gene and its function in most cases often remain unclear.

The application of microbial genetics led to the accumulation of a huge body of knowledge and a continually greater understanding of the nature of genes. The basic research in microbial genetics has not ceased but continues to reveal phenomena important to the understanding of life and its processes as a whole. So, while bacterial genetics paved the way for studying genetic systems other than bacteria, it also ventured to provide solutions for specific industrial, environmental, ecological, pharmaceutical, and other problems. In the early 1970s, microbial genetics itself underwent a revolution with the development of the recombinant DNA (r-DNA) technology. Through these remarkable but straightforward biochemical techniques usually called genetic engineering or gene cloning, the genotype of an organism can be modified in a directed and pre-determined way. The r-DNA technology, in fact, ushered in the era of manipulation of DNA outside the cell, recombination in vitro, and reintroduction of recombinant DNA into a new cell. In this way, novel organisms with characteristics drawn from distant species and genera can be created. For example, human genes can be transferred to a bacterium, and a bacterial gene placed into plants or animal cells. In fact, the glitter of this technology has led to the conversion of hundreds of research laboratories into gene cloning factories, and to the development of a new industry known as bioengineering or biotechnology. Biotechnologists, however, draw heavily from the classical as well as molecular genetics, when it comes to get the required information, and realizing the applications of this technology. Once again this aspect has also not been touched upon in the present treatise, as innumerable information is available elsewhere. In this era of genomics, bacteria figure extensively in genome sequencing projects, adding on to loads of new information. A large number of bacterial genomes have been sequenced, but several gene functions even in the best-studied organisms, such as *Escherichia coli* and *Bacillus subtilis*, remain unknown. Many of these genes do not resemble the other genes characterized in the database, throwing the whole field open for discovering new pathways.

The contents of this book are spread over seven chapters. In Chap. 1, the readers are familiarized with the genetic terminology and some of the basic genetic tools applied to bacteria. Chapter 2 deals with the basic mechanism of mutation, not unique to bacteria, but to which bacteria have made seminal contributions. The next three chapters describe three different pathways through which the inter-bacterial gene transfer is materialized. All these have been essential to generate the genetic variability that has profoundly impacted the bacterial evolution. Chapter 3 describes Conjugation, Chap. 4, Transformation, and Chap. 5 deals with Transduction. Chapter 6 is devoted to the discussion on different aspects of an extra-chromosomal genetic unit, the Plasmid and its Biology. The last chapter describes the Transposable Elements and their contribution to bacterial evolution. A set of important references has been provided and an Index has been appended at the end.

This book intends to initiate the readers into the field of bacterial genetics. Familiarizing them with the tools and techniques of both classical and molecular genetics and exposing them to the strength of bacterial systems in analyzing basic concepts of Genetics, on the one hand, and prepare them to confront newer and newer challenges that bacteria continue to throw at the scientific community.

Acknowledgments

To document comprehensive information on any aspect of bacteria, the invisible, tiny, omnipresent organisms, is a herculean task. These organisms have turned out to be more of a boon for the scientific community, as they can be exploited to gain knowledge on various facets of biology. I have selected the aspects dealing with "Genetics" for this treatise, because this is one area in which they have made their presence strongly felt. Inclusion of all the information that is becoming available through bacteria, in a few hundred pages, however, turned out to be not so easy. I hope that this book will promote better understanding of the subject and ignite young minds to venture into scientific research taking bacteria as model system. While completing this book, information has been drawn from various sources. Therefore

I acknowledge:

- All the authors and researchers who have contributed extensively in this area of science. Their work has not only served as an important resource but also helped format the framework of this book.
- The scientists and investigators, whose work continues to document the usefulness of the bacterial system to answer fundamental questions on life. The immense body of knowledge generated with the help of these tiny organisms has been instrumental in the sharp growth of biological sciences as a whole.
- The students over the years, who posed probing and sometimes uncomfortable questions. Their inquisitiveness has helped keep their perspective in mind.
- My colleagues and friends for their valuable discussions and inputs on the subject. Prof. P. S. Srivastava for critically going through the manuscript.
- The help rendered by Ms. Vandana at various stages in the preparation of this manuscript.

Sheela Srivastava

Contents

About the Author

Sheela Srivastava is currently Professor of Genetics at University of Delhi, South Campus, India, where she has been on the faculty since 1984. With her initial training in Botany with specialization in Genetics at Master's level, she achieved her Ph.D. in Biochemical Genetics of the fungus, *Aspergillus nidulans*. Her post-doctoral stint in bacterial molecular genetics led to her scientific interest focusing on bacteria. Her major area of research is in genetics of metal–microbe interaction, plant growth promoting characteristics of rhizospheric bacteria, peptide antibiotic production by lactic acid bacteria, and metagenomics. Being associated with the Department of Genetics since its inception, she has served as Head of the Department, Dean, Faculty of Interdisciplinary and Applied Sciences, Chairman, Board of Research Studies, and Chairman, Committee of Courses. She has co-authored two books: Understanding Bacteria (Kluwer Academic Publishers, 2003) and Introduction to Bacteria (Vikas Publishing House, 1983) besides co-editing a few volumes. She is currently teaching courses on introductory prokaryotic genetics and advanced courses on bacterial and bacteriophage genetics to post-graduate, M.Phil, and Ph.D. students. Her tag line is: "the most challenging job of a teacher in this field is how to make young students learn genetics".

1 Bacteria and Science of Genetics

Bacteria occupy a unique position in the biological world. Being the simplest of the life-forms, they provide a useful model system to study cell structure, metabolism, structure–function relationships, growth and reproduction, ecology, and many others. It was only in relation to genetics that they found it difficult to acquire an outright acceptability, as they lacked on both the counts that form the basic foundation of classical genetics:

(1) Distinct phenotypic differences between the members of a species.
(2) Controlled hybridization to monitor the transmission of characters from parents to the progeny.

First, they could hardly be seen with the unaided eye and thus distinguishing them by their phenotype was impractical. Second, they were thought not to undergo sexual reproduction, a primary requirement for hybridization experiments. However, with the advent of improved biophysical and biochemical techniques, induced mutagenesis and emergence of biochemical genetics helped clear the way for bacterial entry into the field of genetics. Once having crossed these barriers, they not only paved the way for rapid growth in the area of genetics creating fields like microbial and molecular genetics but also became instrumental in providing study material for devising newer techniques and technologies to understand the life processes better. All this has culminated in ushering in the era of biotechnology and has provided important leads for bringing eukaryotic systems under the realm of molecular biology.

One of the strongest unifying concepts of biology states that species are maintained over a long period of time through a continuity that defines the process of heredity. Implicit in this continuity are the variations that can be seen at two levels: (i) within a species or even the descendants of a single parent. This would mean that any two members of a species even as closely related as brothers and sisters are different enough to be recognized easily; (ii) differences seen amongst the large number of living organisms. These variations have been most critical in driving biological evolution. Bacteria, though the simplest of the life-forms fit very well with both the tenets.

We all know now that the universal unit of inheritance is a gene that corresponds to a stretch of DNA (mainly). Genes are organized on chromosome, and chromosomes are contained within a cellular compartment. Before we discuss the basic tools of bacterial genetics, we shall undertake an overview of the organization of its genetic material.

1.1 Bacterial Nucleoid

Despite the major differences in the cellular organization in Bacteria, Archaea, and Eukarya, they all share the same genetic material, the DNA. Although DNA is remarkably simple,

S. Srivastava, *Genetics of Bacteria*, DOI: 10.1007/978-81-322-1090-0_1, © Springer India 2013

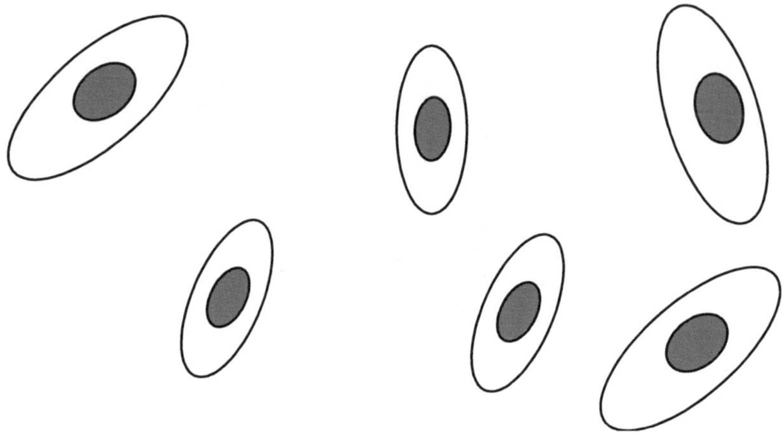

Fig. 1.1 Photomicrograph of bacterial cells stained to locate the nucleoid

it serves as an elegant storage molecule for genetic information. In all living cells, and bacteria are no exception, DNA is inordinately long and requires compaction to get accommodated into a cellular structure. This structure, though organizationally very different in prokaryotes and eukaryotes is known as chromosome. Unlike eukaryotes, bacteria usually contain a single circular chromosome, though linear chromosome and more than one chromosome per cell have been reported. The genetic material is organized in the chromosome in such a way that it is in consonance with DNA replication, chromosome segregation, and gene expression. This is obviously not a simple feat, as the size of the chromosome is generally much longer than the cell in which it resides. In bacteria the chromosome, not withstanding the variation, is generally $\sim$ 1000 times longer than the cell itself.

Structural simplicity has been the hallmark of the prokaryotic organization. This fact was taken to the extent that a unicellular bacterium was considered simply a "metabolic hub" that relies on diffusive processes and does not require a spatial organization. Recent evidence, however, suggest that a bacterial cell in fact, is highly organized. Early investigations, based on the light microscopy of stained cell preparations, showed a compact structure that came to be called as "nucleoid" (Fig. 1.1). Subsequently, electron microscopy of thin sections revealed that although the nucleoid is separate from the overall cytoplasm, it is not bound by a nuclear membrane. This feature became firmly associated with all prokaryotic organizations, and is used as a distinguishing feature to separate them from eukaryotes that possess a true nucleus. The highly condensed nucleoid is located in a pseudo-compartment that is characterized by the absence of the ribosomes (Fig. 1.2a). Another early observation was that some DNA strands were seen extending out in the cytoplasm from an otherwise, what was referred as, a 'tight ball of yarn'. Further studies indicated that these extensions contain most of the transcriptionally active DNA. With refinement in the tools and techniques, it became possible to gently lyse *E. coli* cells, and release the chromosome as a whole. The isolated chromosome appeared as a highly folded structure that contained nearly all the cellular DNA, some proteins, and nascent RNA (Fig. 1.2b).

Electron microscopic images of isolated chromosome of *E. coli* revealed a central core from which $\sim$12–400 topologically independent loops or domains of supercoiling were seen radiating out. The whole structure appeared as an unbroken rosette (Fig. 1.3). This organization was reported to be sensitive to RNase, suggesting that RNA is involved in maintaining the integrity of the core structure. However, this role of RNA is subject to some controversy, and it may just be an isolation-related artifact. The existence of domains of superhelicity has been confirmed by in vivo studies as well, and all the domains were shown to possess the same amount

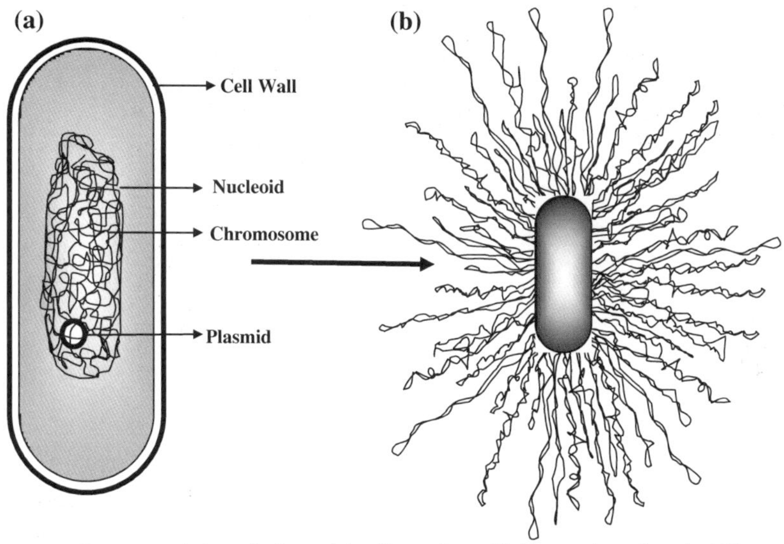

Fig. 1.2 A diagrammatic representation of a bacterial cell carrying a highly condensed nucleoid located in a pseudo-compartment (**a**). A gentle lysis of the cell releases the highly folded chromosome (**b**). *Note* There are no free ends of the chromosome

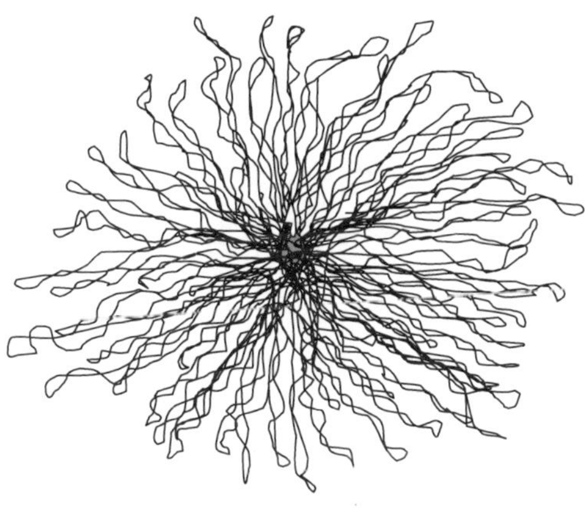

Fig. 1.3 Isolated chromosome of *E. coli* that appears as an unbroken rosette made of independent supercoiled loops

of superhelicity. A single-strand nick abolishes the supercoiling in that particular domain only, leaving the rest of the chromosome unaffected.

The accommodation of the chromosome in a small space within the cell, without jeopardizing the two basic functions: replication and transcription, necessitates its proper folding. To begin with, a cloud of randomly coiled DNA with a diameter of ~ 10 μm is generated in a bacterium, such as *E. coli*. Subdivision of the chromosome into several independent loops of ~ 10 kb provides another level of compaction. Superimposed onto this is the negative supercoiling of the loops that further reduces the diameter of the folded chromosome. Negative supercoiling is the function of DNA topoisomerase that not only leads to compaction but also prevents DNA tangling within the cell. Thus, negative supercoiling is essential for cell survival, as it favors DNA unwinding, enabling many cellular mechanisms to operate. Recent studies have shown that different DNA regions are not only interconnccted but also spatially organized by structure maintenance complex (SMC). SMC acts as a molecular clamp and maintains the chromosome in a state that is compatible with DNA replication and segregation.

However, one of the most important components for DNA condensation is a group of proteins that can alter its shape (compaction) and influence transcription as well. Such proteins have been identified in all the three kingdoms of life. In eukaryotes, these proteins, known as histones, have been studied in great detail, and fairly well understood. Proteins carrying out analogous function in bacteria were also identified and referred to as "histone-like-proteins". This term, however, has become less popular, as we understand these bacterial counterparts better

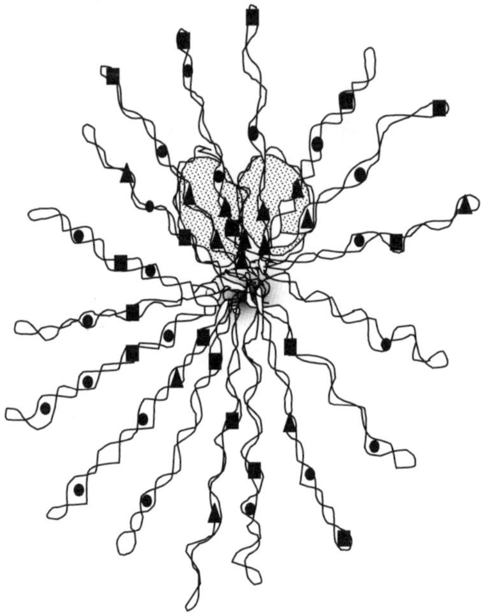

Fig. 1.4 Diagrammatic representation of a supercoiled chromosome along with Nucleoid-associated proteins (NAP-such as Fis and H-NS). Also shown are the transcription complex and RNA polymerase at promoter. Transcription Complex ▲ RNA polymerase at promoter ■ Fis ● H-NS

and better. They are more appropriately referred to as "nucleoid-associated proteins" (NAP), which not only reflects their cellular location but also implies that they may not carry any structural similarity to histones (Fig. 1.4). NAPs consist of numerous and diverse members, many of which are getting identified and added on, but several are not yet fully understood.

Most NAPs possess DNA binding activity and an ability to alter the trajectory of the DNA molecule within the cell by bending, wrapping, or bridging it. Moreover, many can influence transcription in either a positive or negative manner. Some NAPs have a property that allows them to act as boundary elements, i.e., marking the beginning and end of a chromosomal domain loop. NAPs with DNA bridging activity are most apt at carrying out such a function. For example, the NAP, histone-like nucleoid- structuring (H-NS) protein binds to A-T rich regions throughout the genome of *E. coli* and *Salmonella enterica*, but the distribution of binding sites coincides with the domain loop boundaries. Both H-NS and its paralogue StpA can constrain supercoils in DNA and reinforce duplex interwinding in plectonemically supercoiled DNA. Another NAP, "factor for inversion stimulation" (Fis), also binds throughout the genome, and like H-NS is abundant at intergenic sites. Fis is most abundant during early exponential phase of growth, when the looped domains are also in plenty, and is shown to form DNA-protein-DNA bridges. Potential binding sites for Fis, identified throughout the genome of *E. coli*, consist of a consensus sequence, usually 17 bp in length and are AT-rich. Fis binds to these sites as a homodimer and introduces DNA bends. This protein contributes toward many cellular activities, such as transcription, replication, and recombination. HU, another class of NAP, consists of two subunits HUα and HUβ. HU-DNA interactions are non-specific but the protein has a preference for distorted regions in the DNA, such as, bends or four-way junctions. HU plays an important role in recombination, DNA topology management, and gene expression. HU interacts with topoisomerase I that may lead to alteration in superhelicity of DNA, and can be implicated both in the nucleoid structure and gene expression. HU forms multimers consisting of octameric units and has the potential to form spiral filaments onto which a negatively supercoiled DNA may be wrapped around. Besides influencing DNA superhelicity, HU contributes to DNA flexibility by bending the duplex. Such flexibility can facilitate DNA loop formation. Integration host factor (IHF), initially identified as a cofactor in site-specific recombination of phage λ, is a recognized NAP. This protein is related to HU in amino acid sequence, but binds to a well-conserved nucleotide sequence. Binding at the site introduces a U-turn in the DNA, and helps in remodeling of DNA at a local level. IHF influences global transcription, chromosome replication, and serves as a component of site-specific recombination system, as well as affects transposition.

The list of NAPs is growing as more and more are being identified from different bacterial systems. Some other important members are

MukB, Lrp, CbpA, EbfC, Dps, Crp, and many others.

Thus, if we take an overall view, the compact nucleoid occupies the center of a prokaryotic cell, with RNAPs (RNA Polymerases) lying on its periphery, and ribosomes pushed to the edges interacting with the inner membrane. The degree of condensation also depends upon the growth phase, with the highest level of compaction during rapid growth, in comparison to growth under starvation. In the former state, RNAP concentrates in transcriptional foci, whereas under reduced growth (the latter state), RNAP is distributed throughout the chromosome. This suggests that a close correlation exists between nucleoid structure and genome organization. This happens via the distribution of highly expressed genes as their transcription would affect the supercoiling. The association of the whole panoply of NAPs also varies with the growth rate, influencing the topological remodeling of the nucleoid that goes in consonance with the distribution of RNAPs. Differences in the magnitude of nucleoid condensation may actually reflect an adaptation. Bacterial genome size ranges between 200 kb and 13 Mb, and cell size may vary from 0.2 to 750 μm. This will demand a very different degree of compaction in different bacteria. Implicit in this variation is the fact that bacterial nucleoid is highly oriented and organized. What is important is that all the vital functions need to be and are carried out in an ordered manner, and the progressive segregation of the chromosome post replication is enabled. Another interesting observation emerging is that the segregation is not random, and a dedicated mitotic apparatus-like machinery, therefore, is involved. We shall touch upon this subject when we discuss the plasmid partitioning. Although the mode of chromosomal segregation varies significantly from species to species, its mechanism has been intensely investigated and is getting understood.

Thus, when we talk about the genetic material\genes, we are dealing with DNA, as in case of other life-forms. At this stage, the DNA structure is not discussed, as it is expected that readers are well versed with this aspect.

1.2 Genetic Nomenclature

In order to employ bacteria in genetic analysis, we need to first familiarize ourselves with the genetic nomenclature.

Generally speaking, all the strains belonging to a species are compared with a wild-type strain. A wild-type strain is one that shows most of the characters typical of that species. However, a change in a gene compared to wild type may produce a different or a mutant phenotype. The strain itself is then called a mutant. In order to denote the phenotype and/or genotype, some symbols are used. For example, there can be a bacterial strain capable of synthesizing arginine from the components of a mineral salt medium. Such a wild-type strain is phenotypically Arg$^+$ and genotypically *arg*$^+$. In other words, a three-lettered symbol is used and phenotype is written with the first letter capitalized and is not italicized, whereas genotype is indicated by the same symbol written in lower case, italicized letters (underlined). If a mutation makes the strain incapable of synthesizing arginine: its phenotype will be Arg$^-$ and genotype *arg*$^-$ (or simply *arg*).

A phenotype reflects the observable property of a strain. Due to the dearth of simple morphological markers, bacterial phenotypes are based on what they do rather than what they look like. The genotype refers to the genetic composition of a species and is based on identifiable gene functions. One must also realize that arginine biosynthesis, for example, may require more than one gene. In such cases, a fourth letter is added to the gene symbol, each identifying a different gene: *argA*, *argB*, *argC*, etc. Being a part of the genotype this letter is also italicized. For the sake of simplicity, however, the overall genotype and the phenotype for this character will remain to be *arg* and Arg.

Moreover, when we talk of bacteria they are typically haploid and thus carry only a single set of genes. This is an important difference from those eukaryotes that are diploid, and thus carry two sets of alleles on a homologous chromosome pair and will have them in either homozygous or heterozygous combination. In the former, the two copies are similar, *arg*$^+$/*arg*$^+$ or *arg*$^-$/*arg*$^-$, whereas in the latter they are different, *arg*$^+$/*arg*$^-$. Note that the two allele symbols are separated by a slash to indicate that they are physically located on two homologous chromosomes. In other words, in bacteria a gene may exist in different forms (alleles) but each cell will carry only one of them. Thus, bacteria and other haploid eukaryotes (such as many fungi) have become important systems in mutation study, as any mutation (recessive or dominant) introduced will express itself in the very next generation. In comparison, in diploids recessive mutations remain masked by their wild-type allele (*arg*$^+$/*arg*$^-$ will be phenotypically Arg$^+$).

As discussed in detail in another chapter, the basis of generation of an allelic form is mutation. With the advent of induced mutagenesis a large number of mutations could be scored which were initially ascribed a number. Subsequently, when the concerned gene is located (often referred to as locus), the said number is added to the gene symbol. For example, if *arg*-22 mutation was located in *argA*, the latter will now be represented as *argA*22.

Based on their metabolic versatility, bacteria offer a large number of biochemical markers. A marker is equivalent to a genetic locus that has a readily detectable phenotypic characteristic. This may deal with gene involved in amino acid (Arg for example), vitamin, and nucleotide biosynthesis, or those that are involved in the utilization of a whole variety of carbon sources, or resistance to several antibiotics and other toxic compounds as well as various cellular functions. There are some differences in the usage and type of superscript used for other categories of genes. A *lac*$^+$ cell would mean that the strain is capable of metabolizing lactose, and a *lac*$^-$ is one that cannot utilize lactose (note the difference from arg$^+$/arg$^-$). Similarly, a strain resistant to antibiotic ampicillin is denoted as *amp*^{r} (or *bla* standing for β-lactamase, the enzyme responsible for ampicillin breakdown and thus the resistance to the strain) and its sensitive counterpart is labeled as *amp*^{s}. Thus, the choice of the superscript will be determined by the type and state of the marker. It is often useful to identify the gene and its protein product. For example, the RecA protein is the product of *recA* gene. Alternatively, one could call it simply as *recA* product.

1.3 Methods of Genetic Analysis

The term genome denotes all the genes of an organism, which in a bacterium often refers to as genes located on its single chromosome. While talking of genes in their functional form, the term wild type refers to a character/function most prevalent in nature. Such a function is dependent on the species or the strain taken into consideration. A mutant allele, however, will show a changed function. With the functional distinction of the alleles, it is also possible to carry out detailed genetic analyses. This would comprise the dominance test, *cis–trans* complementation test, and recombinational mapping.

1.3.1 Dominance Test

One of such analysis will be to conduct the test of allelism. One of the criteria for this test is to determine the dominance relationship. We have discussed earlier that bacteria being haploid carry only one allele (e.g., a^+ or a) and there is no possibility of having a heterozygous (a$^+$/a) situation to test the dominance. In bacteria, therefore, such a situation is created (for details see Chap. 3 on Conjugation), and depending upon whether expression of a heterozygote is like a$^+$ or a, the dominance of that allele is established.

1.3.2 Complementation Test

Complementation test is carried out to differentiate between allelic and non-allelic mutations, even when the latter affect the same phenotype.

This is based on a simple relationship: while non-allelic mutations always complement, allelic mutations do not.

Such a test will demand two mutant alleles to be tested in a common background. This can lead to two types of arrangements:

(i) both mutant alleles are on one genetic unit (chromosome/plasmid) and the other unit carries the respective wild-type counterparts (*cis* arrangement). For example, two mutant alleles *a1* and *a2* in *cis* configuration will be

$$\frac{\text{a1 a2}}{+\,+}$$

(ii) In the other configuration referred to as *trans* arrangement, both genetic units will contribute one of the two alleles. So the same alleles in *trans* will appear

$$\frac{\text{a1}\ +}{+\ \text{a2}}$$

In *cis* arrangement, all combinations will give wild-type phenotype and thus are non-informative. In *trans*, however, non-alleles will give a wild-type phenotype but the alleles will give a mutant phenotype. Thus, complementation test is always carried out in *trans* configuration. Although simple, it is a very strong test and often resorted to by bacterial geneticists when they have a collection of mutant alleles affecting the same phenotype. In fact, the complementation test with a collection of mutants belonging to the same gene has even helped in the identification of functional subdivisions within a gene, appropriately referred to as complementation groups. In order to carry out this test, two strains each carrying a mutant allele are mated to create a partial diploid, a merediploid, or a heterogenote and the specific phenotype of the same is scored as (+) or (−). In here, a (+) represents complementation and (−) no complementation. This helps in determining the relationship between *n* number of mutant alleles (Table 1.1).

Table 1.1 Results of a *cis–trans* complementation test involving independent mutants of *a* gene controlling the same phenotype

	a_1	a_2	a_3	a_4	a_5	a_n
a_1	−					
a_2	−	−				
a_3	+	+	−			
a_4	+	+	−	−		
a_5	+	+	−	−	−	
a_n	−	−	+	+	+	−

From such a complementation table the following inferences can be drawn:

(i) All the mutants do not behave the same way when put through the test.

(ii) On the basis of their response, they could be classified into a few categories. For example a_1, a_2, and a_n do not complement each other but they do so when it comes to 3, 4, and 5. The same is also true of the group comprising 3, 4, and 5.

(iii) No two mutants of the same type, say for example, a_1a_1 or a_4a_4, complement each other, shown as (−) over a diagonal line.

(iv) This suggests that mutants 1, 2, and *n* comprise one group and 3, 4, and 5 another functional group. Based on this test, these have been referred as cistrons or complementation groups. Cistrons can be given symbols such as A and B to distinguish them. Thus, any two members of a cistron do not complement each other but any two members belonging to different cistrons will complement each other. The cistron is functionally equivalent to a gene.

Some ground rules need to be followed for conducting such a test. These are:

(a) The mutants must be recessive as a dominant mutation will mask another allele and complementation cannot be ascertained.

(b) The mutants must belong to structural gene as regulatory mutations may give a very different response. For example, such a mutant may be *cis*-dominant preventing expression of genes residing on the same chromosome. Similarly, a mutant could be a deletion that removes most of the gene. In both the cases, the mutant will give an all (−) response. Such a response should give a clue that the said mutants need further characterization and cannot be put through a complementation test straightaway.

(c) Occasionally, a mutant may show weak complementation (denoted as ±) with members of one cistron and no complementation with the members of the other cistron. One inference is that it belongs to the latter but inhibits the expression from another gene.

(d) Complementation is only a functional analysis and thus the two members continue to carry the mutant information in future generations.

Complementation is a strong tool not only to understand the functional relationship between alleles but is also useful in determining the number of genes involved in controlling a particular phenotype. The major application of this technique is in the fine structure mapping of a gene. Readers may recall the work of Benzer, who in the early 1950s produced the fine structure map of *rII* gene of phage T4, using complementation as one of the tools.

1.3.2.1 Intragenic Complementation

Many multimeric proteins composed of identical subunits may show intragenic complementation. Although very rare, a few mutants may combine to give a functional or a partially functional protein.

1.3.3 Recombination and Gene Mapping:

Genetic recombination or generalized or homologous recombination is the process through which the genetic material of the two homologous chromosomes is physically exchanged. As the name suggests, it requires strict homology between the participating chromosomes and can happen anywhere throughout the length of the chromosome. The products of this process are two parentals, non-crossovers or unexchanged gene combination, and two recombinants crossovers or exchanged products:

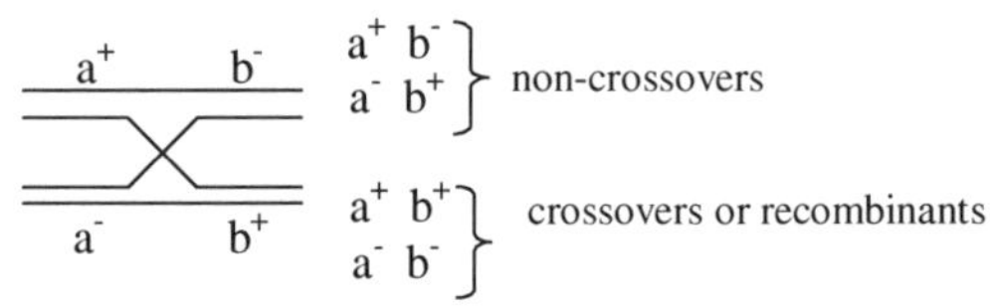

Geneticists from the time of Morgan have used recombination frequency (as percentage) as a measure of distance between the two concerned genes. This can be calculated by the relationship:

$$\frac{\text{Number of recombinant types}}{\text{Number of Parentals} + \text{Number of recombinants}} \times 100.$$

Morgan and his collaborators had proposed that 1 % recombination = 1 map unit distance (now referred to as centiMorgan or cM). In other words, the recombination frequency is proportional to the relative distance between the concerned genes. In order to determine the relative distance along the chromosome one may first have to look at linkage. Linkage is defined as the tendency that allows the genes lying close to each other to segregate together. Thus, the closer the genes, more is the linkage and vice versa. From the Mendelian genetics point of view, genes lying on one chromosome constitute a single linkage group and separate from each other only by recombination. In other words, linkage can be broken by recombination. If we conduct a three-point cross and assume that the genes are equidistantly placed:

I. $\frac{a^+b^+c^-}{a^-b^-c^+}$

One can predict that there will be twice as many ac recombinants as ab, based on the distance. However, this is not always so, and thus recombination frequency may differ between the genes one takes into consideration. If we get 1 % recombination between a-b and 2 % between b-c, two different gene orders can be assumed:

II.

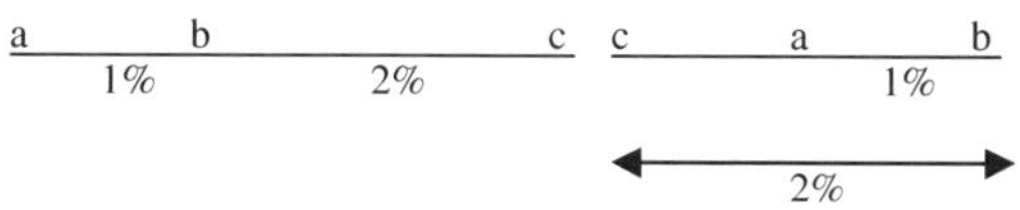

The correct order can be determined from a 2-factor, a × c cross. In situation 1, it should be 3 % whereas in situation 2 it will be 1 %. The two can be easily distinguished.

There is another way by which the relative map location of a gene can be determined without first estimating the distances between them. We shall start with the same cross as depicted in situation I wherein we first ask the question whether *b* is close to *a* or to *c* that is:

III.

We now score for a^+c^+ recombinants and screen them for the possession of b^+ or b^-. If we follow the earlier rule that chances of recombination are proportional to the distance, the possibility of crossing over in the first example in the interval a – b would be far less than in the interval b – c. Thus, most of the a^+ c^+ recombinants will also be b^+. By the same analogy if we analyze the second example, the majority of a^+ c^+ recombinants should be b^-. Once we are able to distinguish between these two arrangements, distances could be determined by 3-point/2-point crosses as shown in situation II.

For some closely placed genes, distances cannot be estimated by simple 2-point crosses. For example if genes *a* and *b* are 5.0 map units, and *b* and *c* are 0.5 map units apart, a simple 2-point *a* × *c* cross cannot tell whether the order is *abc* or *acb*. Theoretically speaking this distance will be 5.5 and 4.5 map units, respectively. Experimentally, however, a – c distance may turn out to be 5.0 ± 0.5 units. A simple method can be employed in such cases. First a cross is conducted:

$$a^+\ b^-\ c^+ \times a^-\ b^+\ c^-$$

Recombinants are now selected which are: a^+ b^+ c^+:

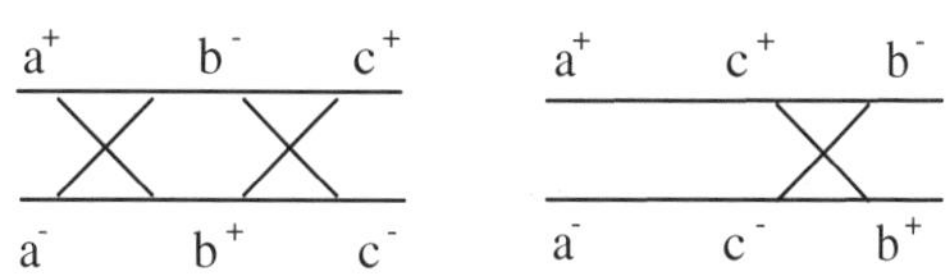

If we follow the two orders, such a recombinant will be formed with two crossovers in the first example and only one cross over in the second example. These two events can be differentiated on the basis of frequency of occurrence. A double crossover will have the expected frequency of the product of two single crossover frequencies (0.05 × 0.005 = 0.025 %), whereas if a single crossover is required it will be the same as b-c recombination frequency of 0.5 %. Once again these two can be easily distinguished.

Mapping exercises are also carried out with the genes that are not very distantly placed. This is once again based on direct relationship between distance and the occurrence of crossing over. Thus, the farther the two genes, more are the chances of crossing over, and sometimes even multiple crossing overs, as shown below:

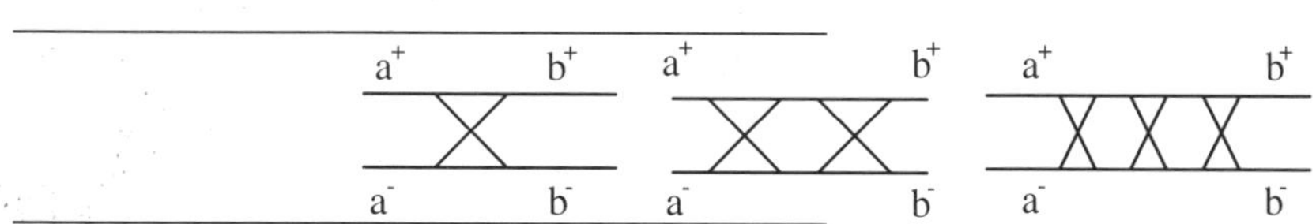

If we compare the three events depicted above, the first and the third will give recombinants but the second will not. Thus, even numbered crossovers do not generate recombinant but odd numbered will produce recombinants between the two genes.

1.3.4 Deletion Mapping

Another mapping procedure available is deletion mapping where a cross is made between a strain carrying a point mutation and the one in which the same gene may have suffered deletion of a different extent. The principle behind this mapping is that only if the mutation and deletion do not overlap, a^+b^+ recombinant, for example, can be obtained, as depicted below:

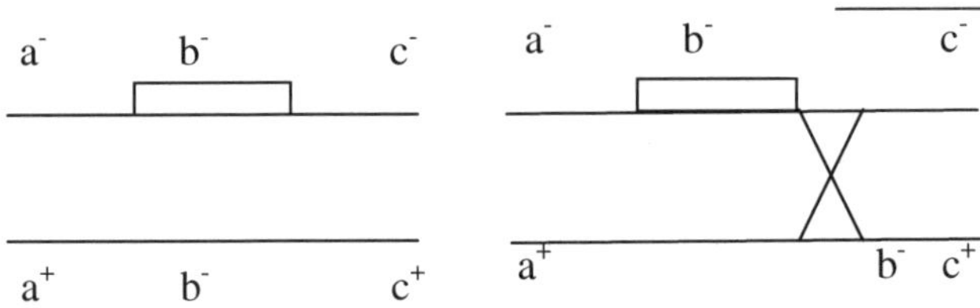

Note A deletion is denoted by showing the region within a box. In the first case, the deletion and mutation are located in the same region and thus will not generate a^+ b^+ recombinant, but in the second, since they do not overlap a^+ b^+ recombinant will be possible.

In fact by crossing two non-overlapping deletions, one could even map them and delineate their boundaries:

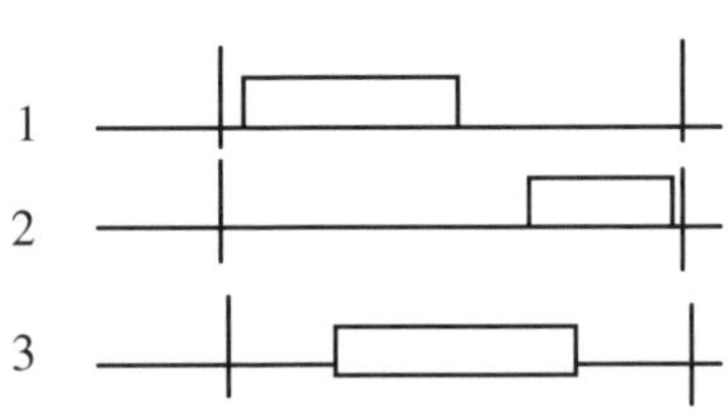

Cross	Recombinant
1 x 2	+
1 x 3	-
2 x 3	-

This analysis is based on a simple yes or no answer, depending on whether wild-type recombinants are obtained. As 1 × 2 produce recombinants their deleted regions must not be overlapping, but both 1 × 3 and 2 × 3 show no recombination suggesting that the deleted region in third stock must be lying in the region common to 1 and 2, as shown below.

Deletion mapping helps in localizing a mutant allele and provides some convenience in making a choice of marker for further recombinational mapping.

1.4 What is a Bacterial Cross?

The whole basis of classical genetics is based on controlled hybridization where parents with desired gene combinations are crossed. Crossing essentially involves the bringing together of desired male and female gametes together for fertilization. The method of such a crossing basically depends upon the mode of sexual reproduction in a species.

When bacteria were first employed for genetic studies they were thought to be reproducing only asexually. However, in 1946, a version of sexual reproduction was discovered in *E. coli* and later in many others. This mode of reproduction is not very prevalent in the bacterial kingdom and therefore, a few parasexual pathways may also be employed by bacteria. Basically, it involves methods by which single-celled organisms exchange their genetic material. We shall study these methods in detail in the subsequent chapters.

Crossing in unicellular organisms like bacteria is far simpler than in higher organisms. Although selection of two parental types on the basis of some selectable differences remains to be the same, it simply involves bringing together two types of cells. With practically no morphological difference in two sexes, the only difference one needs to take care of are the mating type differences such as F^+ and F^- and in the presence or absence of plasmids.

One important consideration in bacterial crosses is the selection of the medium. Such a medium should provide conditions on which only the selected progeny type can grow and in fact should also counter-select the two parental types. This is very crucial as when a mating mixture is plated it contains both the progeny and parental members. Moreover, all possible progeny types are not obtained. Instead what is selected represents only a particular class and the other types are extrapolated on the basis of the reciprocity of the process. To design such a medium, the following must be taken into consideration:

(I) If the two parental type cells carry complementary nutritional requirement, prototrophic recombinants can be selected on minimal agar.

$$a^+\ b^+\ c^+\ d^-\ e^- \quad \times \quad a^-\ b^-\ c^-\ d^+\ e^+$$

$$\downarrow$$

$$a^+\ b^+\ c^+\ d^+\ e^+$$

Note that this medium will also counter-select the two parental strains. Also, only one recombinant type has been considered as only this can grow on minimal agar plates. Although the possibility of other classes of recombinant formation certainly exists, they will be automatically eliminated on a mineral salt medium. At the same time, with a prototrophic recombinant, an all auxotrophic recombinant class ($a^-\ b^-\ c^-\ d^-\ e^-$) will also form out of the same process but will fail to grow on this medium.

(II) In many crosses, the donor (Hfr or male) used, carries prototrophic-markers and the recipient (F^- or female) has counter-auxotrophic markers. In such cases, donor strain is counter-selected by possessing an antibiotic sensitivity or phage sensitivity marker and the recipient strain needs to be resistant for the same. The recipient strain on the other hand is counter-selected by designing a medium that does not fulfill all its nutritional requirements. Recombinants of a particular type are selected on variously supplemented nutritional media with the added antibiotic to which the recipient strain is resistant. Alternatively, the recombinants are selected in the presence of the phage lysate so as to eliminate any contaminating prototrophic phage-sensitive donor cells.

The crosses can be conducted in the following ways:

Liquid mating—The growing population of two strains is mixed and left either static or on a shaker incubator at low speed. This is to facilitate cells' Brownian movement and collision with the desired type of cell. After mixing for different times, cells can be plated on selection agar plates. The ratio in which two strains are to be mixed is also important, as the basis is cell-to-cell contact. This can either be 1:1 or 1:10 or sometimes 10:1 (donor: recipient). In an interrupted mating procedure (as described later), an aliquot of mating mix at different time points is stirred vigorously on a Whirli-mixer and then plated on selection agar.

Spot agar mating—in order to facilitate cell-to-cell contact, some other techniques have also been employed. One of these is known as spot agar mating. In this procedure, after mixing the two strains in an appropriate ratio, the cell suspension ($\sim$200 μl) is spotted on the center of a nutrient agar plate and then incubated overnight. Cells are collected by gentle scrapping, after flooding the plates with 1.0–2.0 ml of saline. The mating mix so collected is stirred vigorously on a vortex mixer and then plated on selection agar plates.

Filter mating—In this method, the two cell types are mixed in an appropriate ratio and filtered through a membrane filter (0.2 μm). The filter paper disk with the cells is placed over the surface of a nutrient agar plate, with the cells side up. The plate is then incubated. After an appropriate time, the filter paper is placed in a tube containing 2.0 ml saline and vortexed to bring the cells in the liquid, and separate them. Cells are then plated on selection agar plates.

Interrupted mating—In many bacteria, like *E. coli*, the donor 'chromosome' is transferred to the recipient in a time-dependent manner. This formed the basis of interrupted mating technique devised by Jacob and Wollman in 1956. Here, the two appropriate parental cell types were mixed, and an aliquot was removed at different times, vortexed in a waring blender to separate the cells or interrupt the mating. In another variation of this technique, cells in a mating mixture could also be diluted several folds. The dilution, however, will not allow any new mating pair to be formed without affecting the already formed pairs. Vortexing, on the other hand, not only disallows the fresh mating pairs but also disrupts the already formed pairs. Cells are then plated on selection agar plates.

1.4.1 Analysis of Cross

From the discussion above, it should become clear that the analysis of progeny of a cross in bacteria will be different from that of eukaryotic system. First, the parental classes will not be obtained. Second, only those recombinant types will be obtained that can be selected. Moreover, in bacterial crosses an intermediate diploid or zygote state is never obtained, instead what is seen are the cells derived from a partial diploid or a merozygote. Thus, a monohybrid cross of $a^+ \times a^-$ will give us a^+: a^- segregation, if both are selected for. Derivation of mapping data is also different as we have seen earlier and shall discuss in detail in the chapters on Conjugation, Transformation, and Transduction.

1.4.2 Isogenic Lines

For any detailed genetic analysis, it is important that two organisms differ only by a single mutation. This is generally referred to be as isogenic lines. If the difference is due to more than one mutation, it becomes difficult to associate a particular phenotype with a specific gene.

1.5 Genetic Exchange in Bacteria

Life evolved on this earth some 3.5 billion years ago. Although it is still difficult to state exactly how the first living organism developed, we do know that they were the simpler prokaryotic life-forms. Facing some of the toughest physical and nutritional conditions, they smartly adapted to the prevailing environment, modulating their metabolic activities such that they slowly changed the earth's atmosphere for other organisms to evolve and survive. This process continued leading to the evolution of new microbial forms. After 3 billion years, during which only microorganisms lived on this earth, the pace of evolution quickened such that over the next 600 million years, a plethora of new organisms, including microorganisms, plants, and animals appeared on the earth. The basic tenet of this evolution was dependent upon the ability of these life-forms to adapt to the prevailing environmental conditions. This ability was and is still drawn from the existing/emerging genetic variability in the population. Variability between different groups and also within the same group of organisms, is an obvious and inescapable feature of life.

In the first 2 billion years of earth's existence, the microorganisms that dominated the scene preferred to reproduce asexually with occasional mutations that apparently would induce genetic variability. Then some 1.5–2.0 billion years ago a remarkable process had evolved: the sexual reproduction, through which an organism would produce genetically variable offspring. The evolution of sexual reproduction is one of the most significant events in the history of life. During this process, the genetic information from two parents is exchanged or reshuffled through the underlying mechanism of recombination. The exchange of genes to produce new combinations of alleles had a profound effect not only on greatly increasing the amount of genetic variation and maintenance of genetic heterogeneity but also allowed for accelerated evolution. Within living cells, the exchange of genetic information (DNA) can occur through an intricately regulated series of enzymatic reactions. Such a sequence rearrangement is what is referred to as genetic recombination, a fundamental process in all cells leading to shuffling of genetic material and generating variability. It will be no exaggeration to state that most of the tremendous diversity of life on earth is a direct result of this genetic reassortment.

Heredity in bacteria is fundamentally similar to heredity in more complex organisms. Thus, although there is an apparent preference for asexual reproduction, bacteria have evolved mechanisms to trade their genetic material. Their success story of this long existence will stand testimony to these mechanisms. The processes of genetic recombination in bacteria differ from that of sexually reproducing higher eukaryotes in two important ways. First, DNA exchange and reproduction are not coupled, and second, the donated genetic material may or may not recombine. Moreover, there is no intervening complete zygote (diploid) formation. Thus, unlike plants and animals which exchange their genetic material every time they reproduce, microbes would do so very rarely, but when they do reshuffle their genes they do it with amazing promiscuity. Undoubtedly, recombination is a critical force enabling bacteria to generate variations that guided the adaptive evolution. It has been shown that the rate of recombination, even among close relatives, is generally close to or rarely more than 10 times the rate of mutation.

Bacteria display enormous levels of divergence in physiology, genome content, and most importantly ecological adaptations. In order to maintain such a level of diversity, bacteria could not have relied solely on the gene pool of its own or the closely related populations. Over millions of years, they could acquire many genes not just from their close relatives, but also from entirely distant members through a process called horizontal gene transfer (HGT). Thus, if a novel gene, conferring selective advantage arises, it can be shared amongst the members of an ecological niche, without an outright concern for relatedness. In the process of horizontal gene transfer, several mechanisms, such as conjugation, transduction, and natural transformation have been implicated. The acquisition of a "foreign gene" became possible not just through a chromosomal segment, but plasmids, phages, transposable elements, and other integrative elements have all contributed toward this gene flux.

While these mechanisms have been known for a long time, novel processes that could extend the existing repertoire of gene exchange methods are continually being identified. For example, temporary cell fusion followed by chromosomal recombination and plasmid exchange, intercellular communications mediated through nanotubes, and the release of membrane vesicles containing chromosomal, plasmid, and phage DNA that can offload these molecules by merging with nearby cells have also been documented. In recent times, yet another novel mechanism of gene transfer combining the features of membrane vesicles trafficking and viral infection, dubbed as gene transfer agents (GTAs) have also been reported. A GTA is a tailed phage—like an entity that contains a random piece of the genome of the producing cell but that does not code for any protein component of the particle itself. The genes encoding the structure of GTAs are contained within the genome of the cell that produces these particles. Therefore, only the occasional particles will contain GTA-encoding genes. These particles are presumably released into the environment by cell lysis and the released particles may transfer the DNA to a recipient cell. Till date, GTAs have been described from some bacteria and archaea but GTA gene cluster has been identified from all taxonomic orders in alphabacteria whose complete genome sequence is available.

In addition to GTAs, several other phage-like elements that carry small fragments of cellular genomic DNA have also been described. All in all these studies indicate that bacteria have explored a variety of known as well as yet to be known mechanisms to achieve one final goal, that is, the reshuffling of their genes so as to add onto the genetic variability that has guided their evolutionary success.

With the exception of plasmids that can successfully implant a donor gene in a replication-proficient form, horizontally transferred-DNA must integrate in the bacterial genome for its persistence and dissemination. The integration is due to homologous recombination, which is often guided by close phylogenetic relationship in more than one ways. First, near relatives will have greater sequence similarity in their shared genes thus enhancing the chances of homologous recombination. Second, close relatives are more likely to share the same habitat than distant ones. This provides a perfect setting for gene exchange. Bacteria have cleverly utilized other ways as well, such as homology facilitated illegitimate recombination (HFIR) that relies on low sequence similarity, or sequence-independent illegitimate recombination events. Thus, bacterial populations keep themselves afloat by acquiring genes of selective advantage that enable them to respond to fluctuating environmental conditions. Recent studies have been directed toward identification of the forces that govern HGT. The role of phylogeny is logically highlighted because the shared evolutionary history will not only favor the gene exchange but also the persistence of genes acquired because of greater genetic compatibility. Some studies have revealed the role of geographical forces providing an alternative structure to HGT by restricting the dispersal. A third possibility is that ecological similarity may shape the gene exchange network by selecting and distributing the adaptive traits or by increasing the physical interactions between the members.

That the HGT has had a major impact on bacterial evolution has been fully corroborated by large-scale genome sequencing. Another fallout of this process has been in the development of clustered regularly interspaced short repeats (CRISPR)/Cas system. Described as an acquired immunity defense system in prokaryotes, it protects the cell against a fresh onslaught by a foreign DNA, such as a phage or a plasmid. CRISPR units comprise a series of short palindromic repeats separated by spacer regions derived from a previous infection/transfer by a phage or a plasmid by HGT, along with many CRISPR-associated (Cas) proteins. There can be an expansion of the spacers, as new spacers derived from foreign DNA (proto-spacers) can be acquired at the leader end of the CRISPR locus (Fig. 1.5). CRISPR loci are transcribed into RNAs, which are cleaved into smaller units and target the complementary sequences of the next round of invading phage- or plasmid-DNA. This may lead to the specific cleavage of the invading DNA, within the proto-spacer at specific sites, thus providing resistance against the same. The *cas* genes encode a large and heterogeneous family of proteins that carry functional domains typical of nucleases, helicases, polymerases, and polynucleotide binding proteins. CRISPR/Cas defense system, therefore, allows a cell to differentiate and tackle invading "foreign DNA" from "self" and survive exposure to invasive elements. This is an important consideration in bacterial evolution, as bacteria may have to protect themselves from foreign genetic elements and preys such as plasmids and viruses. While such a system may help in maintaining the

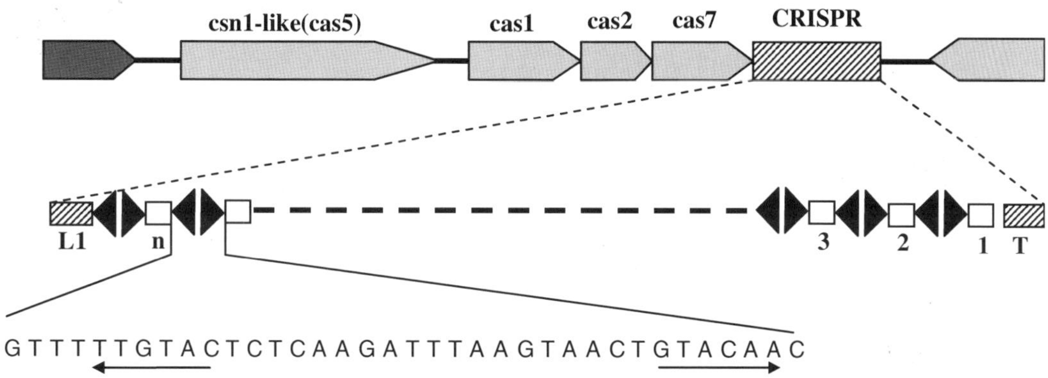

Fig. 1.5 Organization of a representative CRISPR/cas system. The *upper panel* shows its presence at a genomic location. The *middle panel* provides the expanded view of CRISPR, comprising an array of repeats (◀▶) and spacers (□), represents variable numbers (1 to *n*) of these structures. ▨ T represents terminal repeat and L1 is the leader sequence. The *lower panel* provides the sequence of the region carrying the inverted repeats →

genetic integrity, occasional uptake and acquisition of exogenous DNA that may prove advantageous toward an adaptation, is still permitted.

Further Readings

Badrinarayanan A, Lesterlin C, Reyes-Lamotha R, Sherratt D (2012) The *Escherichia coli* SMS complex, MukBEF, shapes nucleoid organization independent of DNA replication. J Bact 194(17):4669–4676

Boto L (2010) Horizontal gene transfer in evolution: facts and challenges. Proc R Soc B 277: 819–827

Dillon SC, Dorman CJ (2010) Bacterial nucleoid-associated proteins, nucleoid structure, and gene expression. Nat Rev Microbiol 8:185–195

Dorman CJ (2009) Nucleoid-associated proteins and bacterial physiology. Adv Appl Microbiol 67:47–64

Garneau JE, Dupis ME, Villion M, Romero DA, Barrangou R, Boyaval P, Fremaux C, Horvath P, Magadan AH, Moineau S (2010) The CRISPR/Cas bacterial immune system cleaves bacteriophage and plasmid DNA. Nature 468:67–71

Gogarten JP, Townsend JP (2005) Horizontal gene transfer, genome innovation, and evolution. Nat Rev Microbiol 3:679–687

Horvath P, Barrangou R (2010) CRISPR/Cas, the immune system of bacteria and Archaea. Science 210:167–170

Lang AS, Zhaxybayeva O, Beaty JT (2012) Gene transfer agents: phage-like elements of genetic exchange. Nat Rev Microbiol 10:472–482

Lederberg J (1960) A view of genetics. Science 131:269–276

Low KB, Porter DD (1978) Modes of gene transfer and recombination in bacteria. Annu Rev Genet 12:249–287

Makarova KS, Haft DH, Barrangou R, Brouns SJJ, Charpantier E, Horvath P, Moineau S, Mojica FJM, Wolf YI, Yakunin AF, van der Oost J, Koonin EV (2011) Evolution and classification of the CRISPR/Cas systems. Nat Rev Microbiol 9:467–477

Ragan MA, Beiko RG (2009) Lateral genetic transfer: open issues. Phil Trans R Soc B 364:2241–2251

Redfield RJ (2001) Do bacteria have sex. Nat Rev Genet 2:634–639

Rocha EPC (2008) The organization of the bacterial genome. Annu Rev Genet 42: 7.1–7.23

Saier MH Jr (2008) The bacterial chromosome. Crit Rev Biochem Mol Biol 43(2):89–134

Stern A, Sorek R (2011) The phage-host arm race: shaping the evolution of microbes. BioEssays 33(1):43–51

Thomas CM, Nielsen KM (2005) Mechanism of, and barriers to horizontal gene transfer between bacteria. Nat Rev Micrbiol 3:711–721

Toro E, Shapiro L (2010) Bacterial chromosome organization and segregation. Cold Spring Harbor Perspective Biol 2:1–31

Wiedenbeck J, Cohan FM (2011) Origin of bacterial diversity through horizontal genetic transfer and adaptation to new ecological niches. FEMS Microbiol Rev 35:957–976

2 Gene Mutation: The Basic Mechanism for Generating Genetic Variability

While the life on this earth may have begun as RNA world, DNA is eventually the genetic material of all living organisms, from bacteria to humans. The only exception comes from some viruses with RNA genome, but then viruses themselves cannot be classified as true organisms. DNA is organized into discrete units, called genes, which not only control the particular characters of an organism but also maintain the genetic continuity as they are passed on from generation to generation.

The genes are highly stable and thus have been responsible for maintaining the stability and continuity of a species. Implicit in this stability is the generation of variation, a critical parameter for biological evolution. One can talk of these variations at two levels:

(1) The variation that takes place in a character and allows the differences to arise in the members of the same species.
(2) The biological variations that accumulate in a population and guide the natural selection and thus the evolution.

2.1 What is Mutation?

While there is no one mechanism to generate such variations, the primary source that brings about a change in a gene is called mutation. As this is likely to change one allelic form into another, it is also called as point mutation. The mutational changes are not only fundamental to biological evolution they are also crucial as no genetic analysis is possible without such allelic variations. In any consideration of a change, a reference point or standard is necessary. In genetics, this point is provided by the connotation, wild type. A wild-type gene may be the most prevalent form found in nature or in the standard laboratory stock. Any change away from the wild type is known as forward mutation and similarly any change toward the wild type is called reverse mutation, reversion or back mutation. Though the term mutation is often applied to both the process as well as the product, an alternative term used to describe the latter is mutant. Another common term used is the 'genetic marker' as the change would identify a gene.

The mutations may arise spontaneously by the action of undefined external or internal factors or may even be induced with the help of agents called mutagens (Fig. 2.1). The frequency of spontaneous mutation, though may vary is very low in comparison to induced ones. In multicellular organisms they may even be differentiated on the basis of the site of origin. For example, those arising in vegetative cells are called somatic and those in reproductive cells as germ-line mutation. The former have a very little chance of transmission as compared to the latter.

Another parameter used to classify mutation is based on the ways in which the mutations are recognized. Accordingly, one class may be called as *morphological mutations*. This class is very common among the highly differentiated

S. Srivastava, *Genetics of Bacteria*, DOI: 10.1007/978-81-322-1090-0_2, © Springer India 2013

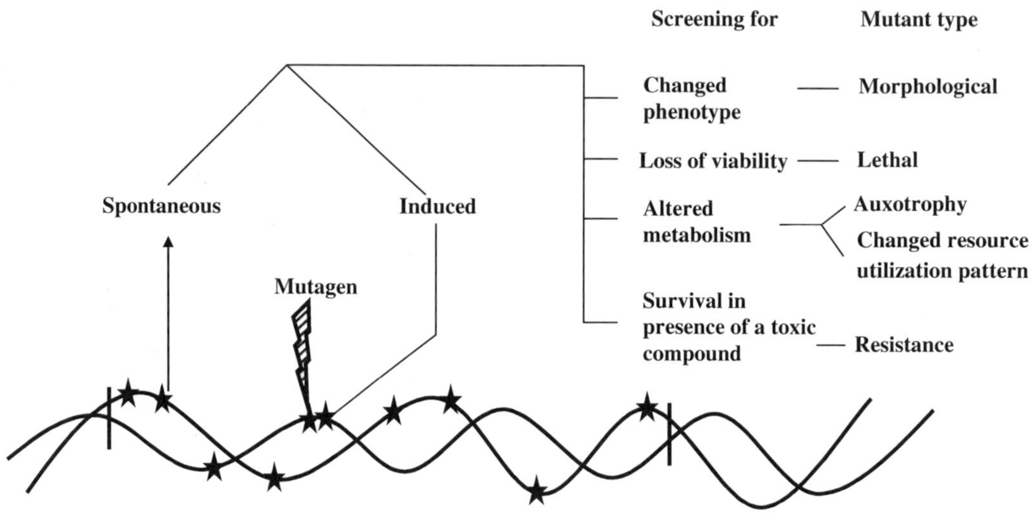

Fig. 2.1 The process of mutation, both spontaneous and induced. *Note* (★) such changes can take place throughout the length of the gene. The *vertical bars* represent a hypothetical length of a gene

higher organisms and appears as distinct changes in the visible characteristics of an organism or phenotype. They are equally rare among the lower organisms, as there are very few perceptible characters. Another class consists of *lethal mutation* recognized by its effect on viability. Such mutations cannot be isolated in haploid micro-organisms as the mutant cells would not survive. In the class of *conditional mutations*, the mutant character is expressed only under certain conditions, called restrictive, and behave absolutely normal under another set of conditions called permissive. Two common types of conditional mutations encountered are temperature sensitive (Ts) and suppressor sensitive (Su) mutants. They constitute a very important class of mutations as they can be easily maintained under permissive conditions and the effect of mutation can be studied by bringing them under restrictive conditions. The class of *biochemical mutations* is identified by the loss or more commonly by a change in some metabolic functions of the cell. Such mutations have been extensively analyzed in microorganisms, as they are not only haploid but also generally prototrophic. Various subclasses of this mutation type can be recognized based on biochemical function in question. The auxotrophic mutations unlike their wild-type prototrophic counterparts fail to grow on a defined minimal medium and require supplementation of essential nutrients. Similarly, many microorganisms possess the capability of utilizing a large number of Carbon sources. By mutation they may lose one such capability. Also, most microorganisms are susceptible to a large number of antimicrobial compounds under wild-type situation, but through mutation they can acquire resistance to them (see Fig. 2.4).

2.2 Why Mutation?

The process of mutation has been going hand-in-hand with life on this planet. All gene functions are liable to undergo mutational change at some time. The mutations can be used for three different purposes: (1) by studying the underlying process of mutation it is possible to identify a gene function, (2) mutations along with recombination permit the genetic dissection of a biological function, and (3) if the changes introduced are useful they can even be exploited.

We shall return to this question later in this chapter and discuss in a comprehensive way, how mutations can be useful?

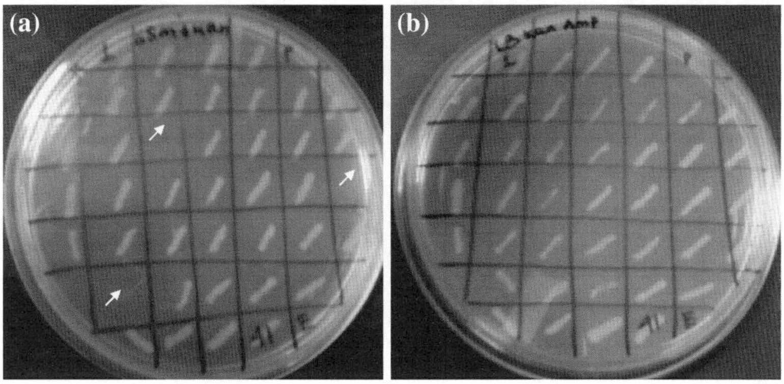

Fig. 2.2 Identification of a mutant by patch replication. Mutagenized colonies are arranged on a master plate (**b**) and are manually patched on selection agar (**a**) in a similar pattern. Absence of growth on selection agar (→) identifies a mutation

2.3 Detection of Mutation

The alterations induced by mutation may not become apparent straightaway. This problem is generally encountered in diploid organisms where two alleles are present and even if one mutates its effect may not be seen in the presence of the wild-type allele unless the mutation is dominant. For such systems, specific tests exist to detect a mutation but will not be dealt with here.

Microbes, being haploid, on the other hand, offer an easy detection system. Depending upon the mutation, they can either be screened or positively selected. The problem of dominance or recessive nature of the mutation does not arise in microbial system as they carry only one copy of an allele. However, if required, situations of partial diploidy can be created, and based on the phenotype, mutant can be classified as dominant or recessive.

2.3.1 Screening

In screening, the mutant colony has to be identified out of the mutagenized (exposed to a mutagen) population. It requires extensive analysis in the form of organizing colonies on a master plate and then patching them or replicating them on selection media on which the mutant is not expected to grow (Fig. 2.2). The respective colony is then picked up from the master plate and its character ascertained. This is normally applied to the isolation of auxotrophic mutation. In patch replication, the mutagenized population is plated on nutrient agar and individual colonies are manually transferred on nutrient agar in a particular pattern (master plate). In order to facilitate the differences between the mutant and non-mutant, the colonies from the latter plate are transferred (patched) in the same pattern on a selective agar plate. These are subsequently compared by their growth response, as described above, and are confirmed for their mutant nature (Fig. 2.2).

In replica plating, a plate carrying 100–200 mutagenized colonies on a nutrient agar plate are transferred on to moist sterile velvet sheet that is held on a replicator by pressing the plate lightly but evenly. Now a plate of selection medium is pressed on to this sheet so that colonies are transferred in the same pattern as the original plate. In both the techniques, the mutant is identified by its failure to grow on selection medium (Fig. 2.3).

2.3.2 Selection

In selection, on the other hand, the medium is designed in such a way that only the mutant is expected to grow, and the non-mutant (wild type)

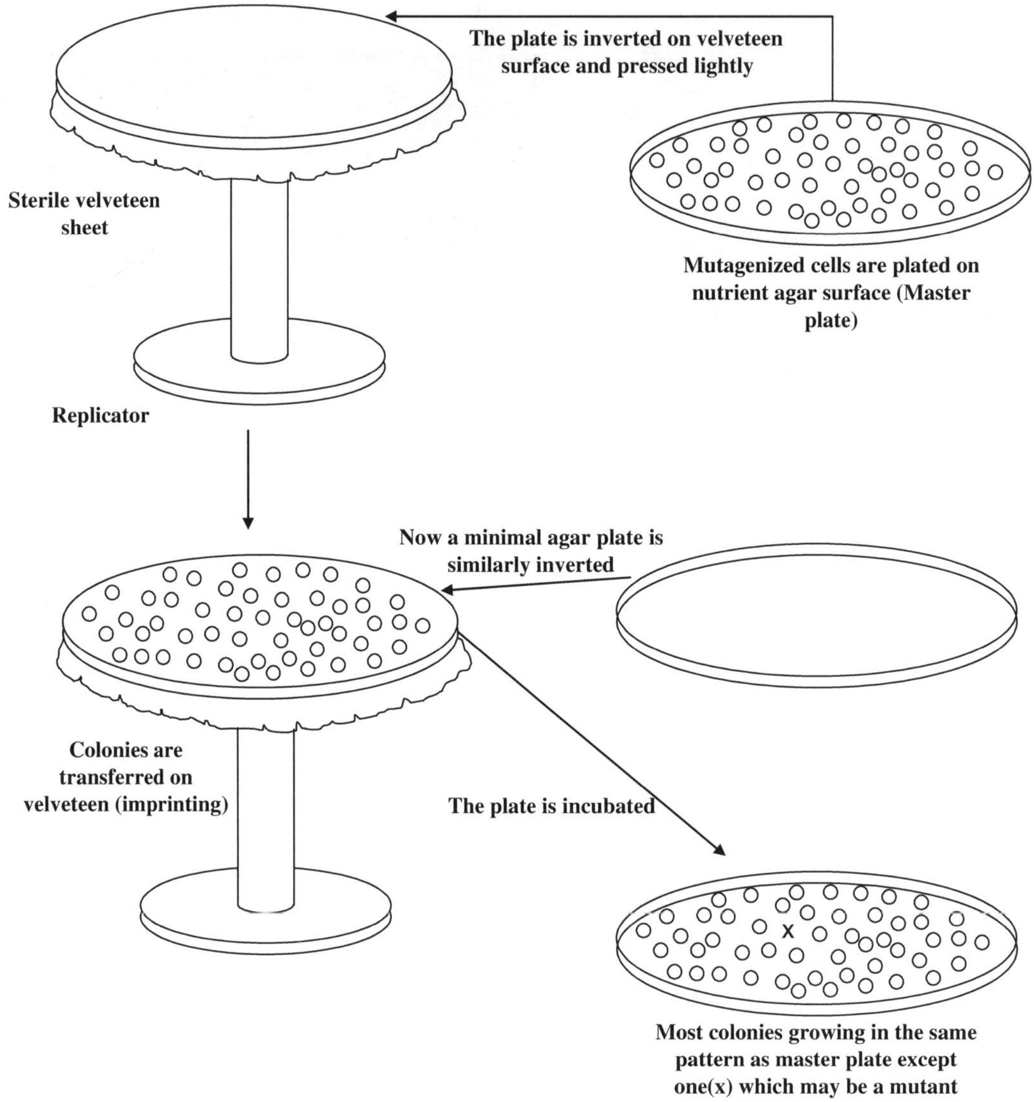

Fig. 2.3 Replica plating for screening of mutants

fails to grow. This method is applicable to the isolation of resistant mutants on a medium containing an anti-microbial compound. On this, the wild type will not grow but the resistant mutants will be able to grow. Another version can be the use of indicator agar on which both wild type and mutants will grow but the latter will show a different phenotype (color) and, therefore, can be easily identified, such as a sugar non-utilizing mutation on MacConkey or EMB agar (Fig. 2.4).

2.4 Characterization of Mutation

Once a colony is suspected to be expressing a changed characteristic, this phenotype is confirmed by reculturing the prospective mutants on a selection medium on a larger streak. The growth performance is rechecked and confirmed. Subsequently, the nature of the introduced mutation is ascertained:

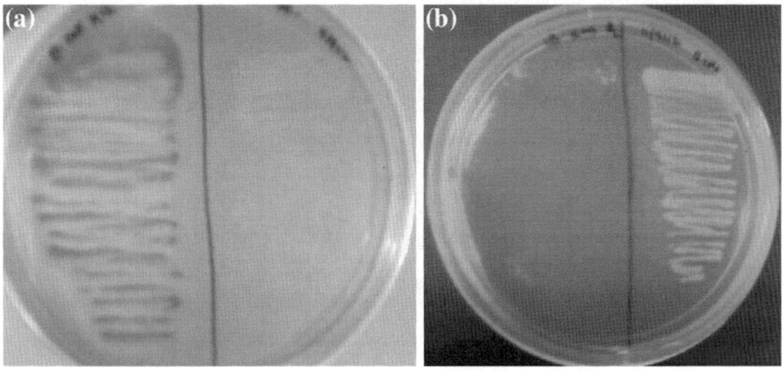

Fig. 2.4 Identification of mutants on (**a**) indicator agar (MacConkey agar) where wild type lac^{+} show *purple/red* color of the colonies and the mutant lac^{-} produces colorless colonies (**b**) in an antibiotic-containing medium, wild type fails to grow but a resistant mutant shows good growth

1. *Morphological mutation.* Any change in colony, shape, size, color, or spores characters for those producing them, can be identified by direct visualization.
2. *Biochemical mutation.* The most common class amongst these are the nutritional or auxotrophic mutations. Such mutants would grow on nutrient agar but fail to grow on minimal agar medium. Once belonging to this category, their requirement will have to be found out (e.g., a vitamin or an amino acid or a base, etc.).

The other class could be with a changed profile of sugar utilization pattern. Many of them could be detected on an indicator agar plate and subsequently rechecked on minimal medium containing that particular sugar as the sole carbon source.

3. *Conditional mutation.* As discussed earlier this is a class of useful mutations where the change is expressed only under certain conditions. Such mutations have contributed significantly toward understanding some essential gene functions in which a mutational change may mean the loss of viability of the organism. Two general classes of such mutations are: (i) the temperature-sensitive (Ts) mutation which behaves as wild type at one or permissive temperature and as mutant at another or restrictive temperature. Often an intermediate phenotype may be seen within this temperature range. It must be clear that the gene is not altered at different temperatures but the product of the gene is inactive at restricted temperature; (ii) the suppressor sensitive mutants exhibit mutant phenotype if the cell does not produce a suppressor gene product ($=sup^{o}$) but the mutant phenotype is masked in the strains carrying a suppressor gene in the genetic background.
4. Resistance mutation—Such mutations can be directly selected for on specially designed media. For example, antibiotic resistant mutants or mutants resistant to other antimicrobial compounds/agents can be selected by plating mutagenized cells directly on nutrient medium containing that antimicrobial compound. Only resistant mutants will grow and sensitive wild types will be eliminated.

2.5 Biochemical Nature of Mutation

Although a mutant is basically identified by a changed phenotype, it is important that the biochemical basis of mutation is understood, as the phenotypic change may not always be easily discernible. As we shall discuss in detail later, mutation is caused by a change in the base sequence of a stretch of DNA that comprises a gene. In fact the mutated sequence is reflected in a changed mRNA base sequence leading to amino acid change in the final product, the protein. Two basic types of changes can occur, generally

Table 2.1 Mutational changes in a hypothetical sequence of a gene

DNA	(i)	TAC	CAC	AAC	TCG	AAA	ACA	ACC	----	ATT	Wild type function
mRNA	(ii)	AUG	GUG	UUG	AGC	UUU	UGU	UGG	----	UAA	
Protein	(iii)	Met	Val	Leu	Ser	Phe	Cys	Trp	----	stop	
	(i)	TAC	**GAC**	AAC	TCG	AAA	ACA	ACC	----	ATT	Substitution
	(ii)	AUG	**CUG**	UUG	AGC	UUU	UGU	UGG	----	UAA	(Transversion)
	(iii)	Met	**Leu**	Leu	Ser	Phe	Cys	Trp	----	stop	mutation
	(i)	TAC	CAC	**AAT**	TCG	AAA	ACA	ACC	----	ATT	Substitution
	(ii)	AUG	GUG	**UUA**	AGC	UUU	UGU	UGG	----	UAA	(Transition)
	(iii)	Met	Val	**Leu**	Ser	Phe	Cys	Trp	----	stop	Wild type
	(i)	TAC	CAC	AAT	**TGG**	AAA	ACA	ACC	----	ATT	Substitution
	(ii)	AUG	GUG	UUG	**ACC**	UUU	UGU	UGG	----	UAA	(Tranversion)
	(iii)	Met	Val	Leu	**Thr**	Phe	Cys	Trp	----		mutation
	(i)	TAC	CAC	AAC	TCG	**GAA**	ACA	ACC	----	ATT	Substitution
	(ii)	AUG	GUG	UUG	AGC	**CUU**	UGU	UGG	----	UAA	(Transition)
	(iii)	Met	Val	Leu	Ser	**Leu**	Cys	Trp	----	stop	mutation
	(i)	TAC	CAC	AAC	TCG	AAA	ACA	**ACT**	----	ATT	Substitution
	(ii)	AUG	GUG	UUG	AGC	UUU	UGU	**UGA**	----	UAA	(Transition) mutation
	(iii)	Met	Val	Leu	Ser	Phe	Cys	**Stop**	----		(truncated protein
	(i)	TAC	**ACA**	**ACT**	**CGA**	**AAA**	**CAA**	**CC**	----	ATT	Deletion
	(ii)	AUG	**UGU**	**UGA**	**GCA**	**UUU**	**GUU**	**GG**	----	UAA	Frame shift mutation
	(iii)	Met	**Cys**	**stop**					----		truncated protein
	(i)	TAC	CAC	AAC	**ATC**	**GAA**	**AAC**	**AAC C**	----	ATT	Insertion
	(ii)	AUG	GUG	UUG	**UAG**	**CUU**	**UUG**	**UUU G**	----	UAA	Frame shift mutation
	(iii)	Met	Val	Leu	**stop**				----		truncated protein
	(i)	TAC	CAC	AAC	**CGA**	**AAA**	**CAA**	**CC**	----	ATT	Deletion
	(ii)	AUG	GUG	UUG	**GCU**	**UUU**	**GUU**	**GG**	----	UAA	Frame shift mutation
	(iii)	Met	Val	Leu	**Ala**	**Phe**	**Val**		----		(changed amino acid)

All changes are shown as (**bold**). (- - -) lines represent the rest of the DNA sequence making a gene
Note Such changes can occur anywhere in the length of the DNA and may result in altered functions

referred to as substitution of a base and insertion or deletion of a base, as described in Table 2.1. From this analysis, we could conclude that:

1. If the alteration is in a single base of a codon it may lead to amino acid substitution.
2. If the alteration is due to deletion or insertion of one or few bases it may lead to a reading frame change resulting into amino acid substitution from that point onwards or loss of one or few amino acids from the protein due to deletion.
3. If the alteration leads to conversion of a sense codon to a stop codon, it may lead to a truncated protein.

Some substitutions may not lead to phenotypic changes and are referred to as silent mutations. In such mutations, the substituted amino acid may either belong to the same class (e.g., by charge) or due to degeneracy of genetic code the mutated codon may still code for the same amino acid (see example above).

Some mutations may produce an effect by which they are referred to as leaky mutation. This is based on the fact that a substituted amino acid may not disrupt the protein function completely. Thus, such a mutant would carry out the gene function partially and will not express a fully mutant phenotype. A mutation may also generate a chain termination codon in the coding sequence of the gene so as to result in a truncated polypeptide lacking the carboxyl terminus.

Another class of mutation may arise from the loss of variable length of the DNA spanning a gene. Such mutations called deletion mutations may give rise to a null phenotype (i.e., no phenotype for that function as compared to the changed phenotype introduced by substitution). Such mutations can be easily differentiated from the point mutation, as the deletion mutations cannot revert whereas a point mutation as a rule must revert to its wild-type counterpart at some time.

2.6 Spontaneous Mutations

As described earlier mutations are random events, and there is no way to know when and in which cells they will occur. Each gene however, is mutable, i.e., it can mutate, and with a definitive probability.

2.6.1 Random, Non-adaptive Mutation

When the microbial mutations began to be isolated it was thought that they arise due to a particular selection pressure, an idea that is finding experimental support in recent years. Isolation of antibiotic resistant mutants in the presence of a particular antibiotic can be cited as an example for the same. In 1943, however, Delbrück and Luria conducted an experiment to demonstrate that mutations are spontaneous and non-adaptive, i.e., they occur even if they do not confer any adaptive advantage, or the change is not guided by the prevailing conditions.

In their experiment, a phage T_1-sensitive (T_1^s), *E. coli* strain was screened for the resistant (T_1^r) mutants. Cultures, each containing the same cell number, were raised in different culture volumes (0.2 ml or small-10 ml or large) and the cells were allowed to grow for the same time. A uniform volume from both (0.2 ml) was plated on agar-medium laden with a high population of phage T_1. The colonies that appeared on each plate were counted and statistically analyzed by applying a fluctuation test. It was observed that although each plate received the same number of cells, the number of resistant colonies was highly disparate when they were derived from small cultures but the same tended to stabilize in each of the large cultures.

Delbrück and Luria explained that this variation in number suggests that mutations are non-adaptive, otherwise the number of T_1^r colonies should have been similar on each plate as the selection pressure (T_1 phage) is common to all of them. The differences can be explained by the randomness of the mutation, i.e., it will depend whether it occurs just before sampling or soon after the plating of the bacteria or any of the divisional stages between these two extremes (Fig. 2.5). In a large culture, the mutant cells tend to get evenly distributed and thus the disparity in their numbers diminishes. This experiment had a significant impetus and led to extensive investigations in the field of microbial genetics.

Delbrück and Luria's experiment got further support when Lederberg and Lederberg devised and applied the technique of replica plating, as described earlier. If the cells from a master plate are replicated on medium spread with phage T_1, only T_1^r colonies which may arise spontaneously due to mutation will grow. Lederberg and Lederberg did show the appearance of T_1^r colonies and demonstrated that their numbers vary from plate to plate suggesting the non-adaptive nature of mutation. However, if the same master plate is replicated onto a number of such selection plates not only the number but also the position of the resistant colonies will remain the same.

2.6.2 Mutation Rate

Although each gene is mutable it mutates with a specific rate in a single generation. Mutation rate reflects the probability by which a gene will be mutated in a single generation. Its measurement has important implications in population

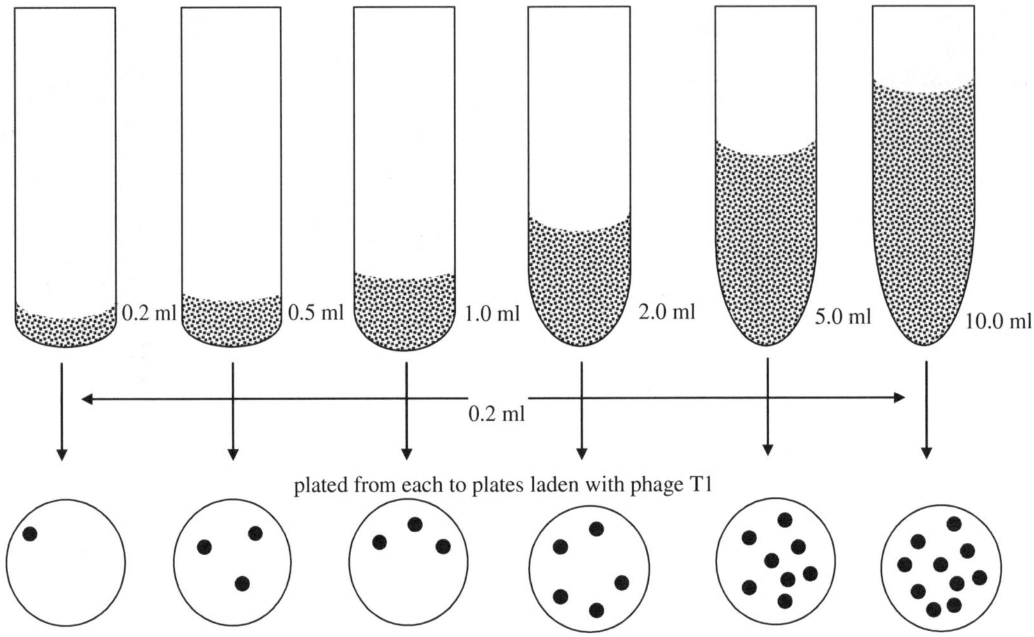

Fig. 2.5 Random non-adaptive nature of mutation as shown by Delbrück and Luria's experiment

genetics, in studies of evolution, and in analyzing the effects of environmental mutagens.

In bacteria, the estimation of mutation rate may get overshadowed by the division/multiplication of the mutant bacterium in a population, and randomness of its occurrence during the growth. If the mutation rate/generation is μ, the probability P_0 of not obtaining any mutants in a culture of N cell can be obtained by Poisson distribution as $e^{-\mu n}$ (though the total number of division required to produce N cells from 1 is $N-1$, the N is large number if it is taken as $N-1 = N$). Thus, $-\mu N - 1\text{n}P_0$, or $\mu = -(1/N)\ 1\text{n}P_0$. If in a fluctuation test, the average number of cells per culture (N) is 5.6×10^8 and in 11 out of 20 samples, no mutants were found, mutation rate can be estimated as $\mu = -(1/5.6 \times 10^8)\ 1\text{n}\ (11/20) = 1.1 \times 10^9$/cell/generation.

2.6.3 Mechanism of Spontaneous Mutation

From the time mutations were observed in natural populations, scientists have tried to reveal the underlying mechanism. In the double helical structure of DNA proposed by Watson and Crick in 1953, some inherent mechanisms of mutation were also suggested by them. Subsequently, a large number of induced mutations from different sources were characterized. All of these could be classified into three categories: (1) errors in DNA replication; (2) spontaneous lesions; and (3) transposable genetic elements.

Essentially, a mutation arises due to a change in the sequence of DNA that comprises a gene. These changes can be brought about in two ways: (1) base substitution—this involves the replacement of a base in a DNA by another base which can occur, either by a purine replacing a purine ($A \rightleftarrows G$) or a pyrimidine being substituted by another pyrimidine ($C \rightleftarrows T$) or when a purine substitutes a pyrimidine or vice versa ($G/A \rightleftarrows T/C$). While the former is known as transition, the latter is referred to as transversion. (2) Frame shift—in these the reading frame of a base sequence shifts due to addition or deletion of one or a few bases. This category of mutation can result in greatly altered proteins (see Table 2.1).

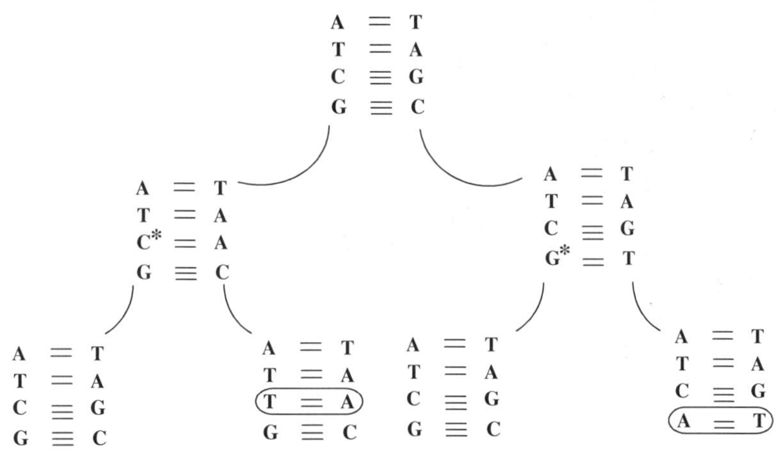

Fig. 2.6 Effect of tautomeric shifts in the generation of transition base substitution that may lead to mutation. The two bases C* and G* are shown in rare imino and enol form, respectively, that may lead to mispairing

2.6.3.1 Errors in DNA Replication

The DNA replication mechanism operates with high fidelity that ensures the continuity of the genetic material from cell-to-cell and generation to generation. However, any error in the incorporation of the correct complementary base will result into alteration and thus may lead to a mutation. The mechanisms that lead to such errors may be largely inherent and can come through in a variety of ways.

The four nitrogenous bases exist in a chemical form called keto form but they have the ability to undergo rare isomerization referred as tautomeric shift to produce molecules with changed hydrogen bonding properties. These are imino or enol forms. For example, in its rare imino form C can pair with A and vice versa and in its rare enol form G bonds with T or vice versa. Such erroneous pairing during replication will lead to base transition (Fig. 2.6). Similarly, if one of the resident bases undergoes such a tautomeric shift it will sponsor the pairing of wrong base causing transitions. Such mismatches can be recognized by DNA polymerase III and excised followed by restoration of the correct pairing. This editing function of the polymerase greatly reduces the observed mutations. Spontaneous transversion cannot arise by replicational error as purine–purine and pyrimidine–pyrimidine pairing can be prevented by steric hinderances.

Replication errors can also lead to frame-shift mutations. In the mid-1960s, Streisinger and his colleagues analyzed the mutations in the lysozyme gene of phage T_4 and observed that frameshift mutations often occurred around the repeat DNA sequences. They proposed a strand slippage model to account for such occurrences. According to this model, during replication, the two strands separate and then re-pair in a way that one or two bases are looped out. When these loops stretch out finally it will lead to addition or deletion of the bases or indel mutations (Fig. 2.7). As is clear from the Fig. 2.7, addition results from the slippage in the growing strand and deletion by this event occurring on the template strand. This model got further boost when Miller and his group were able to assign the mutational hotspots in the *lacI* gene of *E. coli* and correlate them to the distribution of repeat sequences.

2.6.3.2 Deletions

Deletions constitute a sizeable fraction of spontaneous mutations and several of them have been analyzed at the gene and sequence level. Deletions can arise either through the replication error as described above or by bypassing the loop formed by repeat sequences in the inverted orientation during replication. Moreover, unequal crossing over between the repeat sequences located on different DNA molecules or even by recombination between short sequence repeats on the same DNA can also lead to deletions (Fig. 2.8a, b).

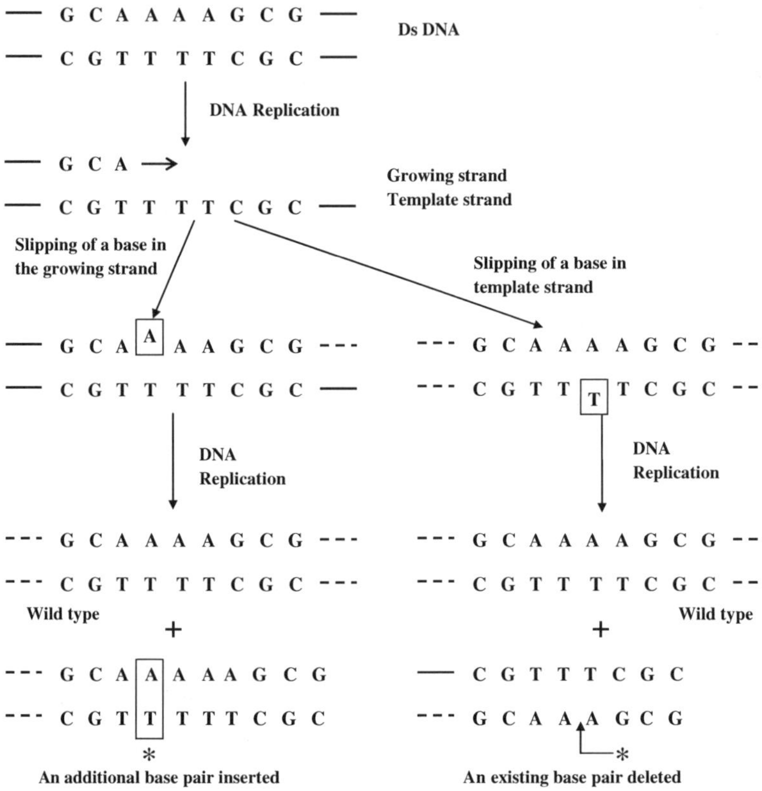

Fig. 2.7 Streisinger strand slippage model to introduce insertions or deletions (*) at a repeat sequence of bases. Such indels (insertion or deletion) may cause frame-shift mutations

2.6.3.3 Lesions

In addition to replication errors, spontaneous lesions can also generate mutations. Two of the most frequent type of lesions comes from depurination and deamination (Fig. 2.9). Depurination is caused by the breakdown of the glycosidic bond between the base and the deoxyribose leading to the loss of a G or A from DNA. Such an apurinic site cannot direct the incorporation of a complementary base and a misincorporation may result in mutation. However, a very specific repair system operates in the cell that could remove and repair such apurinic sites. Thus, the probability of getting a mutation through such an event is always low.

Deamination of cytosine gives rise to uracil, a base otherwise not found in DNA. The cell may contain the enzyme, uracil-DNA-glycosylase, with the capacity to cleave off uracil so generated. Unpaired guanine opposite the removed uracil then directs cytosine incorporation and lesion is repaired. Otherwise, uracil has a base pairing property with adenine and will result into transition of GC → AT base pair.

The analysis of mutational hotspots in several genes suggested the presence of modified base, methylcytosine at those sites. Deamination of methylcytosine generates thymine that cannot be recognized by a repair enzyme as it is a normal base of DNA. The mispaired G–T thus will give rise to a normal G–C and a mutant A–T transition during the next replicational cycle. With this high possibility of introducing a mutation, it explains why methylcytosine is found located at the mutational hotspots within a gene.

2.6.3.4 Oxidation of Bases

The reactive forms of oxygen such as peroxides and free radicals are often produced as by-products of metabolism. These can react with bases and alter their pairing behavior (Fig. 2.9). For example, 8-OxoG produced by alteration of G can mistakenly pair with A causing G–C → T–A transversions.

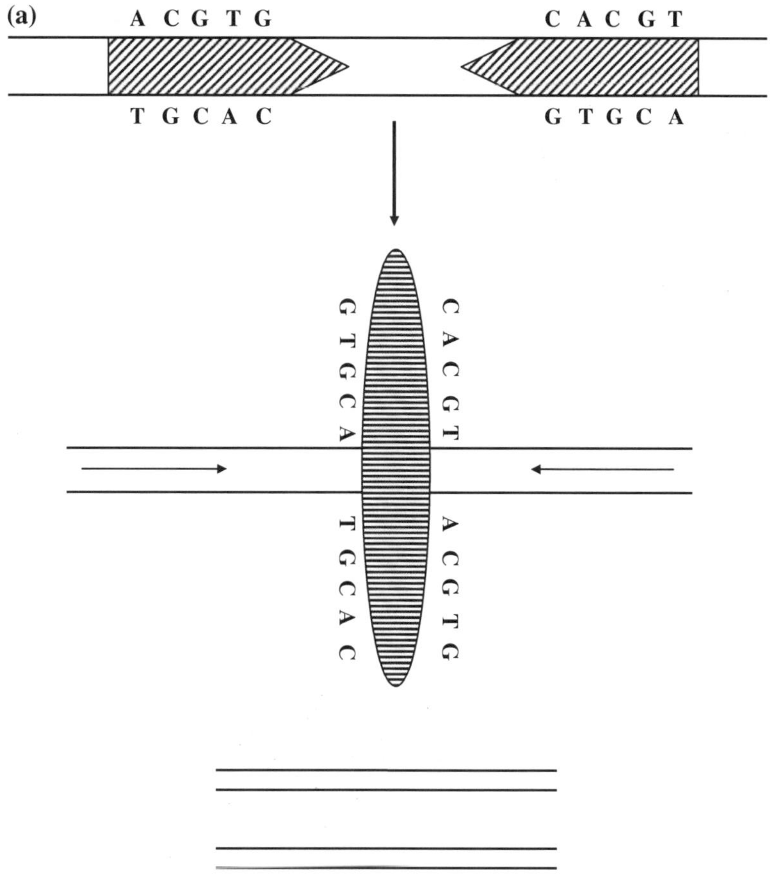

Fig. 2.8a Generation of deletion by bypassing the loop formed by repeat sequences in inverted orientation, at the time of replication. *Note* Acridine dyes use this pathway by stabilizing the loop.

2.6.3.5 Transposable Elements

Since the discovery of mutation, a lot of debate has been generated as to the nature of the agents responsible for causing mutations. With the better understanding of the biological systems, many of these agents/processes were found to be occurring within the cell and to which genetic material is constantly exposed to. As far as external agents are concerned, they range from particulate radiations emitted from radioactive materials, and cosmic radiations, to several other harmful compounds. In 1960s, a new class of elements, first discovered about a decade ago in the plant, maize by Barbara McClintock, was also reported in bacteria. These elements were identified by their strong mutagenic properties and came to be known as transposable or mobile genetic elements. Many of these elements are resident members of bacterial chromosome, plasmids, and phages and have the ability to move from one site to another in a genome, and once it moves within a gene, it may cause a mutation. We shall discuss these elements in detail elsewhere but it would be worthwhile to mention here that a large number of spontaneous mutations in bacteria could be traced to the mobility and insertion of these elements in a functional gene. We shall discuss about these elements in detail in Chap. 7.

2.7 Induced Mutations

Until 1927, though a number of mutations in different organisms were observed, analyzed and studied, any serious study on them was obliterated by the fact that the rate of spontaneous mutation was generally very low at practically all loci. In 1927, Müller demonstrated the induction of mutations in *Drosophila* by X-ray

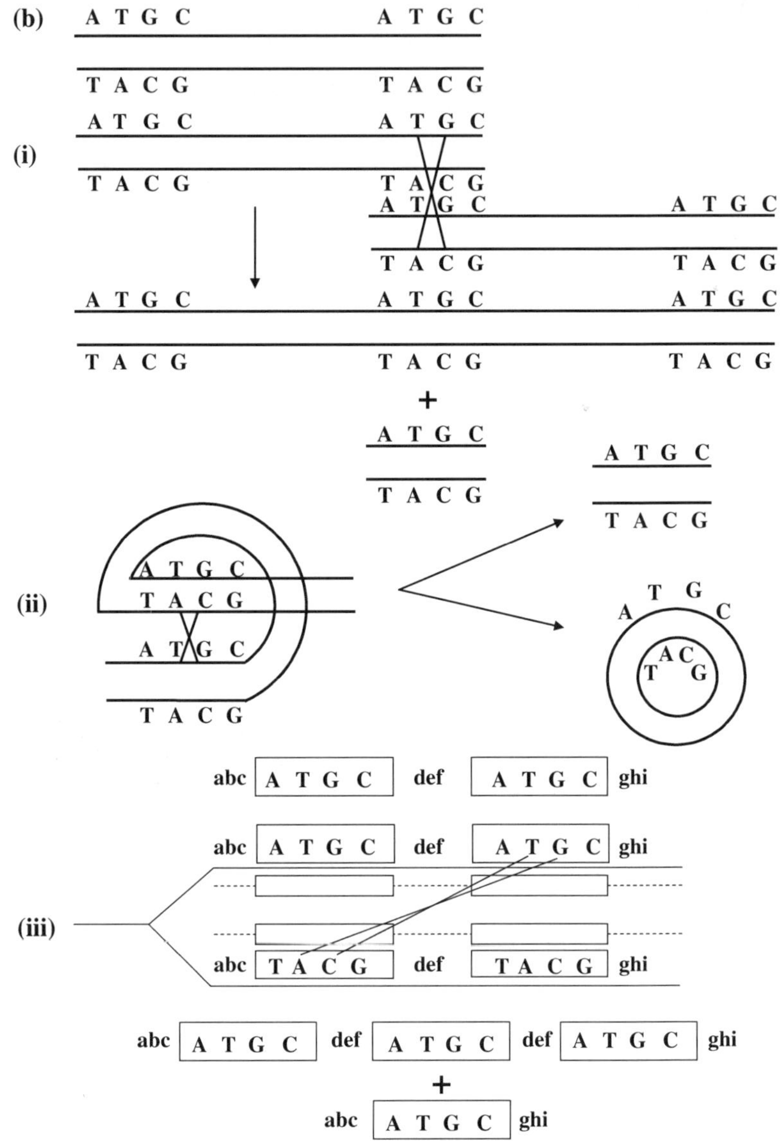

Fig. 2.8b Generation of spontaneous deletions using repeat DNA sequences represented by *tetranucleotide stretch.* (i) Unequal crossing over between repeat sequences on different DNA. (ii) Recombinational deletion involving repeat sequences on the same molecule. (iii) Unequal sister chromatid exchange. *Note* In both (i) and (iii), a duplication will also result

which was soon followed by a similar demonstration by Stadler (1928) in Barley. Since then, a large number of agents causing mutations, called mutagens, have been reported. Studies with the help of mutagens enabled the scientists to understand the underlying mechanism of mutation and the cellular responses to overcome these mutagenic effects. Generally speaking, such mutants help in unraveling the gene function and thus are of prime importance in any genetic analysis. These mutagens have been classified into two broad groups: physical mutagens and chemical mutagens.

2.7.1 Physical Mutagens

They basically consist of high energy electromagnetic radiations of a variety of wavelengths. They have been classified into ionizing and nonionizing radiations based on the pathway through which they operate. The visible light

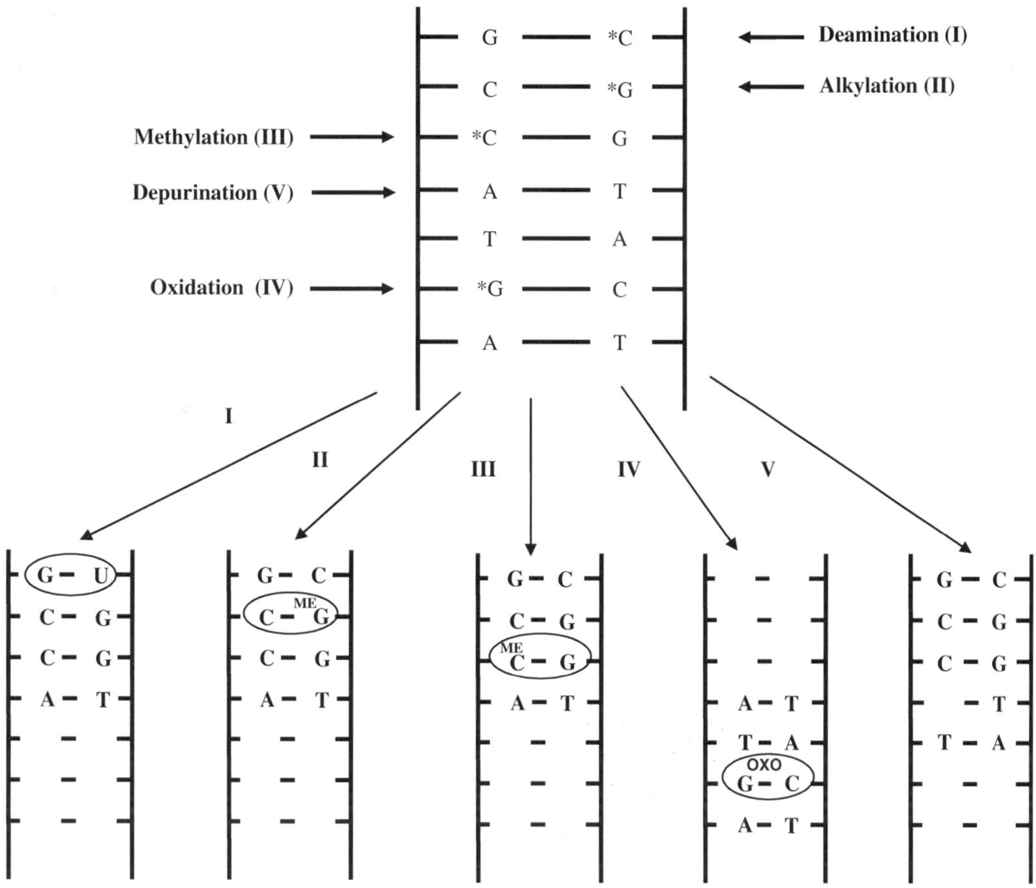

Fig. 2.9 Generalized pathways of DNA damage. A hypothetical stretch of DNA has been drawn to show what can happen due to damages that occur independently. Changes are shown as * in the original DNA duplex and resulting modified bases that will lead to mismatch are encircled. In (V), depurination leads to a gap due to removal of a purine. If not repaired, these modifications can lead to mutation

perceived by our eyes comprises a very small component of these radiations (Fig. 2.10). As the wavelength decreases, the energy levels become stronger and thus more penetrating. X-rays discovered in the nineteenth century by Röntgen have a wavelength ranging from about 10 to ~0.1 Å, and thus have a strong penetrating ability. This formed the basis, as to why they were employed not only in diagnostic medical procedures but also as mutagen.

During penetration, X-rays collide with other atoms and eject electrons which can hit other atoms releasing further electrons. This chain reaction brings about a change in the stable state of the atom to reactive ionic state and earns the name of ionizing radiation to X-rays and several other radiation types. To this category also belong alpha, beta, and gamma rays emitted from radioactive sources such as radium and ^{60}Co, etc. Alternatively, they bring about ionization of other molecules which may produce reactive forms and that in turn can react with DNA to induce mutations. Because of their high penetrating power and ionization ability, they often induce chromosomal breaks and the chromosomal aberrations.

Ultra-violet radiation (UV) though considerably milder than X-ray, is also effective in inducing mutations. Although its mutagenic properties were first demonstrated in *Drosophila,* the detailed mechanism has been worked out using microbial systems. It has a relatively long

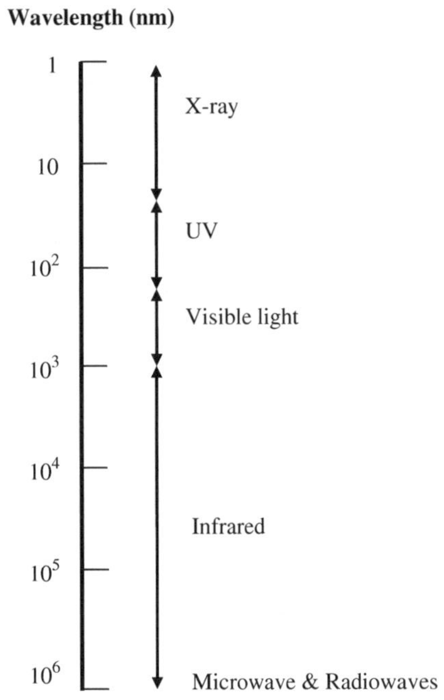

Fig. 2.10 The electromagnetic spectrum of different types of radiations

wavelength, is less penetrating, and non-ionizing and affects only those compounds that absorb it directly. Based on the wavelength and its consequent effectivity, UV irradiation has been classified into different types (Table 2.2).

In a cell, compounds having organic ring structures such as purines and pyrimidines are able to absorb UV in the range of 254 nm wavelength, a wavelength that is most mutagenic.

The major effect of UV is thymine dimerization, which may occur between the two adjacent thymidine bases in one DNA strand or it may form cross-links across the two strands. These are also known as cyclobutane pyrimidine dimers (CPD), because a cyclobutane ring is formed between C atom 5 and 6 of the adjacent thymines (Fig. 2.11). Such dimers distort the helix and interfere with proper replication. Alternatively, cytosine can be hydrated and this brings about a change in its pairing behavior. Moreover, UV can also react with other compounds, for example, media components and indirectly affect the organism. All cells possess dedicated or more generalized repair pathways, as discussed below.

2.7.2 Chemical Mutagens

Although irradiation provided a way by which a gene can be modified to probe its structure and function, they were not always very refined in their action as they may induce gross aberrations or mutations and that too by their direct or indirect effect. A large number of chemical compounds possessing mutagenic properties were discovered right from the time of World War II, and with the advent of man-made chemicals, this list is ever growing, several of them possessing carcinogenic properties as well.

Some of these mutagenic compounds are nitrous acid, nitrogen and sulfur mustard, ethyl- and methyl-methane sulfonate, diazomethane, and others. They are often classified on the basis of their mode of action, as described below (Table 2.3), and each one of them invokes specific repair pathway. We shall discuss both the aspects under DNA damage and repair.

Certain chemicals produce interstrand cross-links in DNA that obviously hinder DNA replication. These are mitomycin C and dimethyl psoralen. The latter compound needs to be activated by exposure to 360 nm wavelength light, and thus provides a practical control on its activity.

2.7.3 Base Analogs

Base analogs are chemical compounds that structurally resemble the normal DNA bases, and thus, can be easily incorporated into newly synthesizing DNA strand. This event can be mutagenic as analogs often pair with the wrong base leading to transitional base substitutions. 2-Aminopurine (2-AP) is an analog of adenine but sometimes pairs with cytosine causing AT → GC transitions. Similarly, 5-bromouracil (5-BU), which resembles thymine can pair with guanine causing similar TA → CG transitions.

Table 2.2 Classification scheme of ultraviolet radiations

S.No.	Type	Wavelength (nm)
1	NUV (near UV)	400 to 200
2	UVA (long wavelength or black light)	400 to 320
3	UVB (medium wavelength)	320 to 280
4	UVC (short wave germicidal)	Below 280
5	FUV or VUV (for or vacuum UV)	200 to 10
6	EUV or XUV (extreme or deep UV)	31 to 1

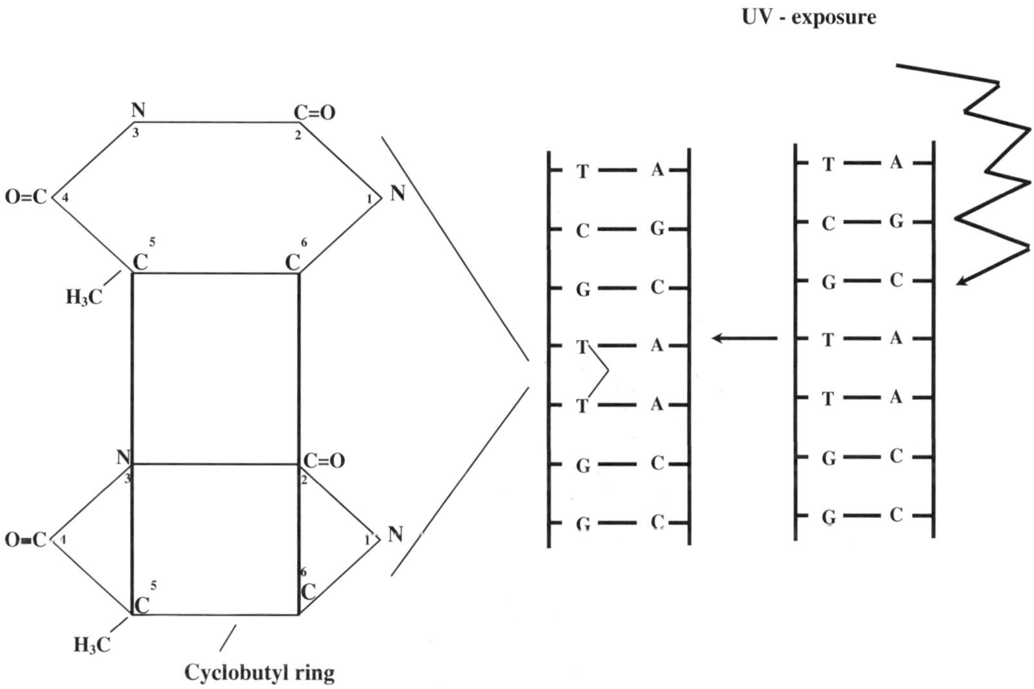

Fig. 2.11 Exposure to UV radiation may lead to pyrimidine dimers in DNA (Thymine dimers or cyclobutane pyrimidine dimers) that are implicated in mutagenesis. The *thick line* shows as to how the two thymines are linked

Table 2.3 Some chemical compounds that function as effective mutagens and their mode of action

Name of the compound	Class of compound	Mode of action
5-Bromouracil (5-BU)	Base analogs	Introduce transitions AT⇄GC
2-Amino purine (2-AP)		
Nitrous acid, hydroxylamine	Modifying the bases leading to a changed base pairing	Deaminates A, C leading to transition GC⇄AT
Ethylmethane/ methylmethane sulfonate (EES/EMS)	-do-	Adds ethyl group to all the bases, promotes GC⇄AT transition
Nitrosoguanidine (NG)	-do-	Adds alkyl group, with an addition to oxygen at position 6 of G leads to AT⇄GC
Acriflavine, acridine orange, ethidium bromide	Acridine dyes	Intercalates between the bases leads to addition/ deletion of base causing frameshift mutation

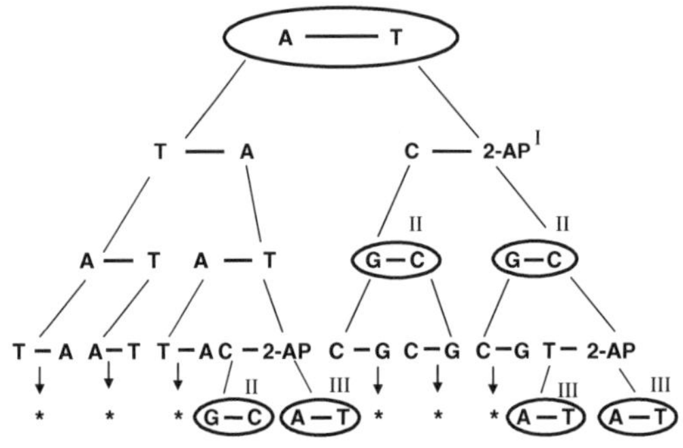

Fig. 2.12 Effect of a base analog (2-AP, 2-aminopurine, an analog of adenine). While its incorporation in replicating DNA (I) can lead to base substitution, based on its changed pairing behavior (II), it can also reverse the effect later (III), based on tautomeric shift. The base pair, where, there is no incorporation of analog will continue in a normal way (*). *Note* All DNA base pairs have not been depicted at the same level of replication

Base analogs, to begin with can also mispair and then resort to correct pairing. For example 2-AP may initially pair with cytosine and in subsequent replications may pair with thymine. This will also cause GC → AT transitions (Fig. 2.12).

2.7.4 Frameshift Mutagens

Many members of the acridine dye family, such as 9-aminoacridine, proflavine, and ethidium bromide are planar molecules having capability to stack between bases of DNA. This intercalation forces the two bases to move apart and the two strands slip with respect to each other. Such a slippage is more likely to occur when a base pair in DNA is repeated, for example, a stretch of AT or GC base pairs. If the dye is intercalated in a DNA strand before replication, the newly synthesized strand may slip and incorporate an extra base (addition). However, if it finds a place in newly synthesized strand, the strand might slip backward during subsequent replication leaving out a base pair (deletion). We have discussed this pathway elsewhere also while explaining spontaneous mutations. Acridine dyes are known to stabilize these slippage loops, facilitating frame-shift mutation (see Fig. 2.7).

2.8 DNA Damage and Repair Pathway

Different agents damage DNA in different ways and the cells can invoke different repair pathways to overcome such damages (Fig. 2.13). Such repair pathways may range from being highly specific to more generalized in nature, as we shall describe below. It should also become clear that under limits, a cell can manage all the damages caused by different mutagens, it is only when the damage is too heavy, and repair itself becomes error-prone that mutations may arise. This is a very important consideration, as it helps the cell to maintain the stability of its genetic material; changes are allowed only rarely. The majority of mutations arise by changing the base sequence of a codon that codes for a different amino acid. Such mutations are known as missense mutations. However, some may lead to the conversion of the sense codon to a chain termination codon and are called nonsense mutations. The effect of such mutations on the protein product will depend upon the location of this mutated codon in the DNA. If the mutation takes place in one of the early codons, it may generate a truncated protein or no protein at all.

2.8.1 Deamination

One of the most common types of DNA damage is deamination. Although deamination can often occur spontaneously, especially at higher temperature, many chemicals called deaminating agents can also remove amino groups from DNA bases. This effect is highly mutagenic as the deaminated base tends to mispair. Adenine for example is converted to hypoxanthine that pairs with cytosine instead of thymine. Similarly, deamination of cytosine produces uracil that has pairing specificity with adenine instead of guanine. Such agents depending upon their action will cause GC → AT and/or AT → GC transitions. Nitrous acid is a classical example of this group of mutagens.

Hydroxylamine specifically reacts with cytosine and thus causes GC → AT transitions. This is a potent in vitro mutagen and works best on purified DNA or viruses. The efficiency of mutagenesis increases if the mutagen-treated DNA is transferred to the cells lacking the uracil-N-glycosylase repair system, as described below.

Bisulfite also deaminates cytosine, but for this cytosine must be present in single-stranded DNA. Bisulfite is thus the mutagen of choice for site-directed mutagenesis as described later.

2.8.1.1 Deamination Repair

Special enzymes referred to as DNA-glycosylases have evolved to remove the deaminated bases. These enzymes break the glycosyl bond between the damaged base and the sugar in the nucleotide. A unique DNA-glycosylase exists for each type of deaminated base. Efficient removal of uracil arising from deamination of cytosine is important for preventing C to T transitions. The enzyme, uracil DNA-glycosylase (Ung), therefore, excises the unwanted uracil from the genome by singling out the rare U–A or U–G base pairs in the predominantly normal background of T–A or C–G base pairs. Ung, however, does not distinguish U from T, while it lies within the DNA helix, instead discrimination occurs in a transient pyrimidine-sieving pocket in the enzyme. This pocket solves the central kinetic problem in efficient base flipping, i.e., the extremely rapid entry of the base into DNA duplex. After the removal of deaminated base, nucleases called AP endonucleases cleave the sugar-phosphate bond next to the missing base. Two such sites can be envisaged: one from where a pyrimidine (C, T, or U) has been removed (an apyrimidinic site) or from where a purine (A or G) has been excised (an apurinic site). After the cleavage, the free 3′-OH end so created is used as a primer for repair DNA synthesis, whereas 5′ end is degraded exonucleolytically. Thus, a whole region around the deaminated bases is resynthesized by incorporating the correct bases (Fig. 2.13a).

2.8.2 Very Short Patch (VSP) Repair of Deaminated Methylcytosine

DNA of most organisms may contain 5-methylcytosine in place of cytosine at some specific sites. Such a modification is brought about by the transfer of methyl group at 5th position in the pyrimidine ring by the enzyme methyltransferase. The presence of 5-methylcytosine may not only protect the DNA from restriction endonuclease digestion but is also implicated in regulating gene expression. However, it may also provide a mutational hotspot as the deamination of 5-methylcytosine will give rise to thymidine which being a normal base of DNA

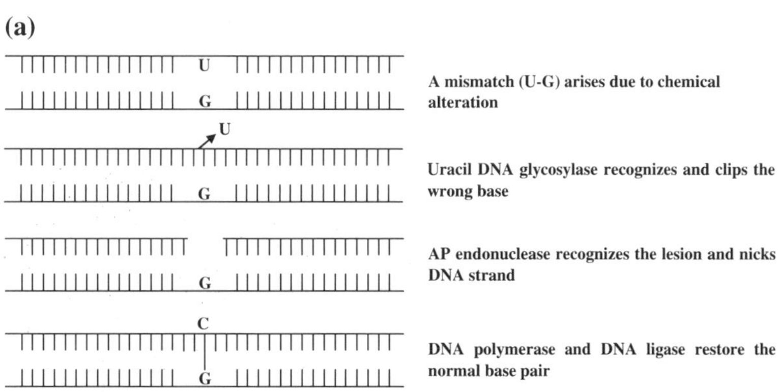

Fig. 2.13a Base excision repair: Uracil generated through base modification is recognized and excised by uracil DNA glycosylase, the apyrimidinic site created is nicked by AP endonuclease, gap is filled and ligated to restore normal base pairing

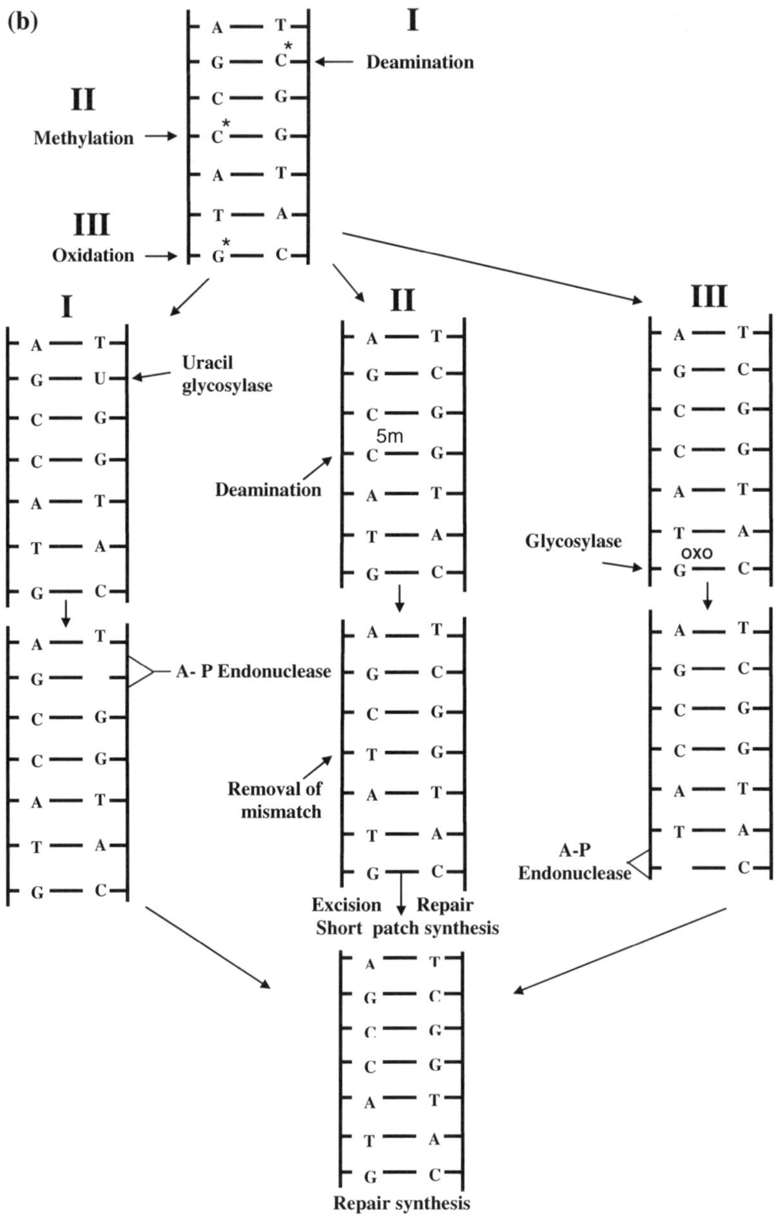

Fig. 2.13b Some of the common DNA damage repair pathways. Note that apurination site will be repaired by cleavage with AP endonuclease and subsequent gap filling

cannot be recognized by any glycosylase. As this thymidine would form opposite guanine, it can still be repaired by methyl-directed mismatch repair, as discussed later in the chapter.

In a G–T mismatch created by replicational error, the wrong base can be identified because of its unmethylated status. However, a similar mismatch produced due to deamination may not occur in a newly synthesized strand. The repair, therefore, becomes random, and if it occurs in the wrong strand a GC → AT transition will result.

In *E. coli* K-12, most of the 5-methylcytosine is found at the second C in the sequence 5′CCNGG3′/3′GGNCC5′, in which N is AT or TA. The methylation of the second C is brought about by an enzyme DNA cytosine-methylase (Dcm). Since this has a high mutation potential, *E. coli* has evolved a repair system that specially removes T whenever it appears as a TG mismatch in this sequence. Because a very small region or "patch" around wrong T is removed and resynthesized, this is also known as "very short patch" repair. The Vsr endonuclease involved in this process is coded by *vsr* gene, which lies immediately downstream to the gene coding for Dcm methylase. This ensures the specific mismatch repair as and when the damage occurs.

2.8.3 Reactive Oxygen Damage

A cell may produce various reactive forms of oxygen either during normal cellular metabolism or during detoxification of some chemicals or even due to environmental factors such as UV-irradiation and some herbicides. These include superoxides, hydroxyl radicals, and hydrogen peroxide, which are all very damaging to DNA. Thus, all aerobic organisms have evolved mechanisms to overcome the damaging effect. This may include the induction of enzymes such as catalase, peroxidase, and superoxide dismutase, which help to remove the reactive forms of O_2. Some repair enzymes may also take care of such damages caused to DNA.

One of the most potent lesions caused by reactive oxygen molecule is the oxidized base 7, 8-dihydro-8-oxoguanine (8-oxoG or Go) having a pairing preference for adenine. Several *mut* mutants, showing a higher than normal rate of spontaneous mutation, have been identified in *E. coli.* While some of these genes such as *mutS*, *mutL*, and *mutH* contribute to methyl-directed mismatch repair system and *mutD* that encodes the editing function of DNA Pol III, the products of three *mut* genes, *mutM*, *mutT*, and *mutY* provide, protection against oxidative damages such as 8-oxoG-mediated mutation. The MutM is an N-glycosylase as well as AP-endonuclease. So it not only cleaves off 8-oxoG but also cuts the apurinated strand. Subsequently, exonuclease degradation and DNA Pol I-mediated resynthesis repairs the strand. The gene *mutM* is a part of a regulon that is induced during oxidative stress. The MutY enzyme is also a specific N-glycosylase but it removes misincorporated A rather than 8-oxoG. The overproduction of MutM completely suppresses the mutation phenotype of *mutY*. The action of MutT enzyme is very different as it prevents the incorporation of 8-oxoG in DNA. The reactive forms of oxygen not only can oxidize guanine in DNA but also the base dGTP to 8-oxo-dGTP. DNA polymerase cannot differentiate between dGTP and 8-oxo-dGTP as a result the latter can also be incorporated in DNA. The MutT enzyme is a phosphatase that converts 8-oxo dGTP to 9-oxo dGMP and prevents its usage during DNA synthesis (Fig. 2.13b).

2.8.4 Alkylation

A large number of chemical compounds known as alkylating agents function by alkylating the bases or phosphate. These include ethylmethane sulfonates (EMS; nitrogen mustard gas), methylmethane sulfonates (MMS), and N-methyl-N′-nitro-N-nitrosoguanidine (nitrosoguanidine NTG, or MNNG). These agents add alkyl groups to reactive groups of the bases. For example, N^7 of guanine and N^3 of adenine can be alkylated by EMS or MMS to N^7-methylguanine or N^3-methyladenine. This change completely alters the pairing function and causes major distortions in the helix. NTG, on the other hand, can also attack other atoms in the rings, such as O^6 of guanine, which is converted to O^6-methylguanine and similarly of O^4 of thymine that forms O^4-methylthymine. These alterations may not bring about any major distortion in the helix and thus may go unnoticed by the general repair pathway but their mispairing would induce mutations.

The repair of alkylated bases is brought about by an adaptive response that consists of two major pathways. One involves specific N-glycosylases, in that the alkylated base is removed, an AP endonuclease cuts the DNA strand, the exonuclease degrades the cut strand and DNA Pol I repair synthesizes to fill the gap. Three genes *alkA*, *alkB*, and *aidB* have been identified as components of this pathway, of which *alkA* product is an N-glycosylase that removes many alkylated bases.

The second pathway employs special proteins called methyltransferases, which directly picks up the alkyl group and in the process get themselves inactivated. One such gene identified in *E. coli* is *ada*. The Ada protein is a broad-functioning methyltransferase that also regulates the transcription of other genes of the adaptive responses pathway.

2.8.5 UV-Irradiation

UV irradiation is one of the major natural sources that can damage DNA. As this type of irradiation forms a part of visible sunlight, every organism that is exposed to sun rays must possess a repair pathway. UV effect is based on the fact that DNA bases because of their conjugated ring structure can absorb light in the UV wavelengths as described earlier. The absorbed photons energize the bases causing their bonds to react with other nearby atoms resulting in additional bonds either between the bases or between bases and sugars.

One of the most common types of such an abnormal linkage is pyrimidine dimerization. In one type, 5-C and 6-C of two adjacent pyrimidines are joined to form a cyclobutane ring (see Fig. 2.11), and in the other type, the linkage involves 4-C and 6-C of adjacent pyrimidines to form a 6-4 lesion.

Two major pathways, photoreactivation, and excision repair are employed to repair UV-induced damages (Fig. 2.14). However, presently we shall discuss only the former as it is specific for dimer repair, whereas the latter may be used for more general purposes as well.

The photoreactivation or "light repair" is so called as it occurs in the presence of visible light. Though first discovered in a bacterium, *Streptomyces griseus* is now known to exist in most organisms except placental mammals like humans. The enzyme involved is photolyase that contains a reduced flavin adenine dinucleotide group ($FADH_2$) that can absorb light of wavelengths between 350 and 500 nm. Thus, the enzyme binds to the dimerized bases, and uses the energy derived from absorption of light to separate the bases. It is believed that binding of this enzyme to pyrimidine dimers can also make the defects easily recognizable by the excision or dark repair system.

Specific N-glycosylases that recognize and remove pyrimidine dimers have also been reported. Their mode of action is same as that during repair of the deaminated and alkylated bases, as described before.

2.9 General Repair Mechanisms

Not all repair mechanisms are so specific to a certain type of DNA damage. Some broad-based repair mechanisms have also evolved that recognize a whole range of damages such as distortions in DNA structure arising due to improper base pairing. These may generally come through the action of base analogs, some acridine dyes causing frame shift mutation or even some alkylating agents. Base analogs have a two-way action, and thus often can reverse the effect of a change that they themselves bring about. For example, 2-aminopurine (2-AP), an analog of A, as described earlier, can bring about A–T to G–C transition, but by mispairing with C it can also reverse the effect, resulting in G–C to A–T transition. Similarly, frame shift mutations caused by deletion or addition of one or few bases can be reverted by subsequent addition or deletion events, respectively. The reversion would, however, be possible only if the two events occur within a short distance from each other, and the two changes do not affect the function of the derived protein.

2.9.1 Methyl-Directed Mismatch Repair

This repair system involves three *mut* genes, *mutS*, *mutL*, and *mutH*, and *dam*. The *dam* gene

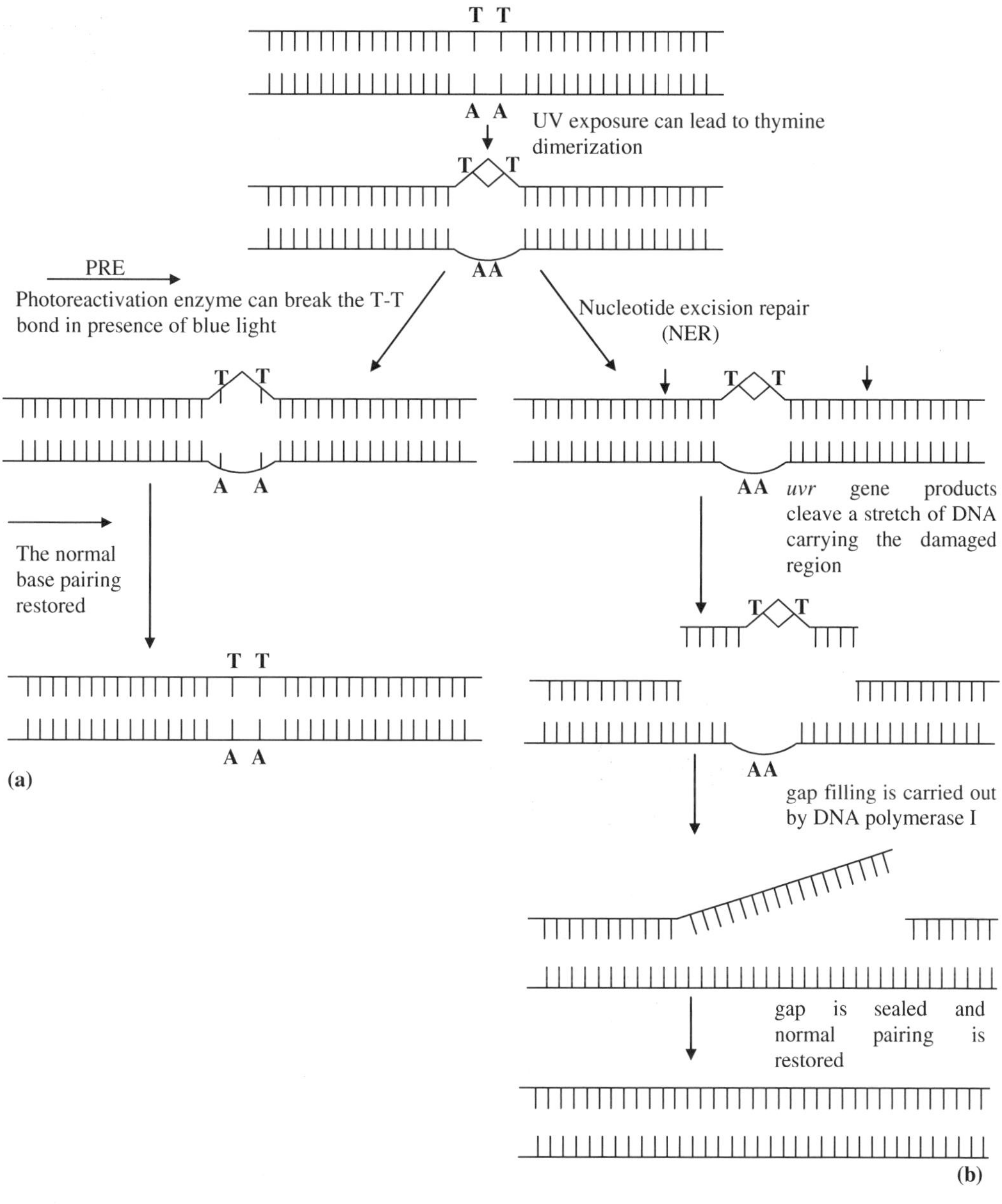

Fig. 2.14 DNA damage repair pathways that may operate after UV exposure. (**a**) Light or photoreactivation repair. (**b**) Dark or nucleotide excision repair (NER) *Note* All types of bulky lesions leading to distortion of the helix are also subjected to NER

codes for Dam-methylase that methylates adenine in the sequence GATC/CTAG. Since this methylation takes place only after new DNA synthesis, it helps in identifying the newly synthesized strand, which will be temporarily unmethylated. If a mismatched base is introduced through a replicational error, the *mut* gene products will initiate the repair only in the newly synthesized strand. First, the MutS protein recognizes and binds to the altered (mismatched) site. Then two molecules of MutH and one MutL bind to MutS, and the DNA slides through the complex in both directions to form a loop held by these proteins. When one of the MutH in the

Fig. 2.15 Methyl directed mismatch repair

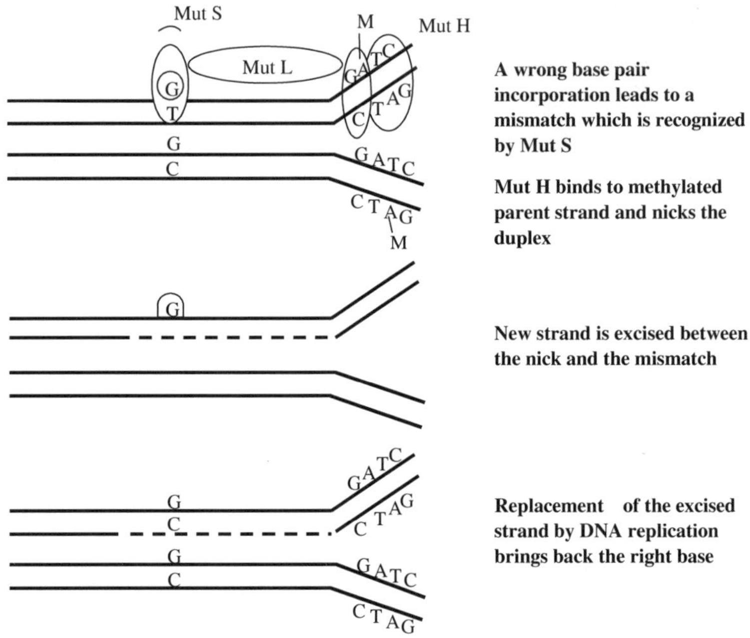

complex reaches a hemi-methylated GATC/CTAG sequence in either the 5′ or 3′ direction, it cuts the unmethylated strand, which is subsequently degraded. Excision of a part of DNA stretch is accomplished by a cooperation between UvrD (helicase II) that unwinds DNA from the nick toward the mismatch, and a single-strand-specific exonuclease. DNA polymerase III then resynthesizes the removed portion of the new strand, based on the information provided by the template strand, and the same is finally sealed by DNA ligase (Fig. 2.15). The role of MutL in this repair process is not very clear though as a part of the complex, it is proposed to coordinate the activities of the other two proteins. MutS and MutL also participate in another repair pathway called very short patch repair involving methylated cytosine. As described earlier, deamination of methylcytosine generates T, leading to a T–G mismatch. MutS protein binds to such a T that attracts Vsr endonuclease to the mismatch site. The MutL protein then recruits UvrD helicase and exonuclease to locally degrade the strand carrying the mismatch. The gap is then repaired by DNA Pol III.

2.9.2 Nucleotide Excision Repair

One of the most important general repair pathways is nucleotide excision repair (see Fig. 2.14). This is highly efficient, fairly non-specific, and universal in existence. Like mismatch repair system, it recognizes the distortions caused in the DNA helix rather than the chemical structure of damage itself. However, it also functions against major distortions like those caused by UV-irradiation. In this repair system, the stretch carrying the damaged DNA is actually excised and subsequently repaired.

Several *E. coli* genes have been identified which participate in excision repair. For example, *uvrA*, *uvrB*, and *uvrC* are specifically involved whereas *uvrD*, *polA*, and *lig* have more generalized contributions. Mutations in any one of these leads to higher sensitivity to UV exposure. The products of *uvrA*, *uvrB*, and *uvrC* interact to form UvrABC endonuclease (or UvrABC excinuclease). In fact, two copies of the UvrA protein and one copy of UvrB form a complex that binds non-specifically to DNA even if it is not damaged. This complex then surveys the DNA up and down till it finds a distortion, when it stops. At this point, UvrB protein binds to the damaged region allowing UvrA protein to leave. The latter is then replaced by UvrC. This causes UvrB to cut the DNA $\sim$4 nt 3′ to the damage and UvrC cuts the DNA $\sim$7 nt 5′ of the damage. UvrD protein being a helicase helps in releasing the cut damaged DNA, DNA Pol I resynthesizes, and ligase seals the ends to restore the DNA molecule. The *uvr* genes are induced by DNA damage thus falling in a class of genes called *din*, which includes *recF*, *recA*, *umuC*, and *umuD*. Many *din* genes constitute a part of SOS regulon also, as described below.

2.9.3 SOS Inducible Repair

Very often the cells may face emergency situation when they are either inflicted with heavy DNA damages due to various environmental factors, mutagens or even infections. This requires a very dedicated response in which the survival is the most important factor. Such a response is logically inducible and often termed as SOS repair. Several genes are induced after DNA damage but all are not induced at the same time, and to the same level. This phased out response is important in providing the required protection to the cell. To begin with *uvrA*, *uvrB*, and *uvrD* are induced. Their protein products along with UvrC endonuclease catalyze the nucleotide excision repair pathway, as described earlier. As a second line of defense, *recA* and other homologous recombination pathway genes get induced, which bring about the DNA lesion repair, through recombination pathway. Some other genes involved, for example, *sfiA*, a division inhibitor is induced to cause the transient delay of cell division so that the damage could be repaired, and others allow the cells to replicate past the damaged region of DNA. As mentioned above, many *din* genes operate under SOS response, which rescue the cells from severe DNA damages. Finally, if the damage is too heavy and all these repair systems are not able to handle the same, DNA repair polymerase V, encoded by *umuC* and *umuD*, is induced.

The SOS operon remains repressed by LexA repressor that under the normal conditions, binds to a consensus sequence (SOS Box) in the promoter and prevents the expression of the concerned genes. In the event of the DNA damage, the LexA repressor gets inactivated thus allowing the expression of these genes (Fig. 2.16). The functional repressor is a dimer each with a dimerization domain and a DNA binding domain. The repressor inactivation is an autocleavage step that leads to dissociation of repressor from the DNA. When excision repair or post-replication repair is invoked single-stranded DNA tends to accumulate in the cell. Such single-stranded DNA provides the target for RecA binding which in turn stimulates the

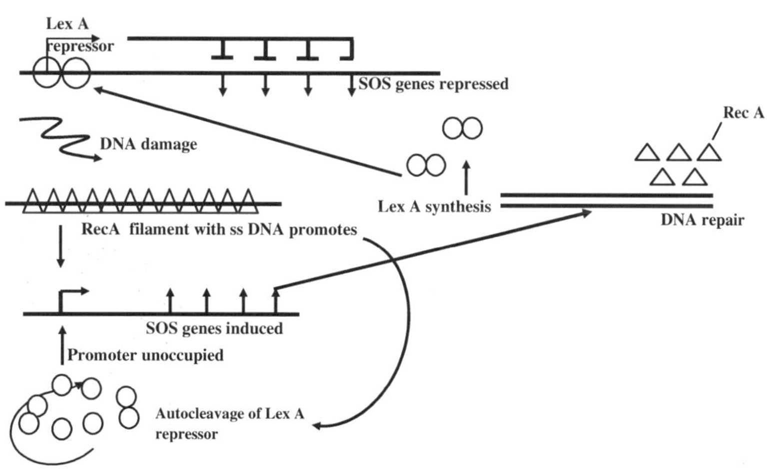

Fig. 2.16 Induction of SOS Repair Pathway

autocleavage of LexA. RecA is a multifunctional protein also involved in the process of genetic recombination, as we shall discuss later, which in here acts as a co-protease.

At least one of the repair systems (*umuDC*) under the SOS control appears to be error-prone and thus often causes mutation (also known as mutagenic repair). The first indication of an inducible repair was provided by Jean Weigle when he showed that irradiated phage λ survived better on irradiated *E. coli* in comparison to unirradiated hosts. This was called Weigle reactivation (or W-reactivation). Also there were more mutants among survivors if the bacterial host was irradiated. By the same analogy this was termed as Weigle (or W) mutagenesis.

Both nucleotide excision and recombination repair do not seem to be mistake-prone but the one coded by *umuDC* is also autocleaved, and it is the cleaved form (UmuD′) which is mutagenic. However, larger quantities of RecA would be required for cleavage of UmuD than that for LexA. This suggests that mutagenic repair is used as a last resort by the cell when the damage is too heavy to be taken care of by other pathways.

It has been proposed that UmuC and UmuD′2 (DNA polymerase V) proteins allow the replication to proceed across the lesion. The process known as replicative bypass or trans-lesion synthesis, permits the replication and cell division but damaged bases may not pair properly which can lead to mutation. Normally, the replicative DNA polymerase will stall at the regions of DNA damage. This is because the damaged base cannot pair properly and thus the entering base would be taken as wrong insertion by the editing component of the polymerase, thus stopping the same. The editing function is assigned to β–protein that forms a circular clamp holding DNA polymerase on to DNA. By comparison, the UmuC, and UmuD′2 proteins may function to form a more relaxed clamp that allows the DNA polymerase to read through the damage with the likelihood of incorporating a wrong base. Such a polymerase may also lack a 3′–5′ proof reading activity and thus, may provide for a much higher rate of incorporation of a wrong base. However, this polymerase can add only a few nucleotides before falling off. This will prevent the formation of a long stretches of DNA containing mismatches. This type of mechanism seems to be universally present amongst bacteria.

2.9.4 Post-replication or Recombination Repair

In some repair pathway, the damage is not repaired instead bypassed by allowing the replication over this region. Thus, the damage is gradually diluted out. These mechanisms may be more appropriately called as DNA-damage tolerating systems. Both DNA polymerases, IV and V are able to mediate trans-lesion DNA synthesis. Polymerase IV, called DinB (coded by *dinB*) and polymerase V, also known as Umu D2′C complex is coded by *umuDC* operon. Both can bypass the DNA damage that may otherwise block the replication by DNA polymerase III. They are prone to make mistakes and thus have been implicated in mutagenesis. One such damage-tolerating mechanism is post-replication repair as it operates after replication has proceeded past the damage through a recombinational step.

The consequences of DNA damage occurring in the lagging strand is very different from those happening in the leading strand. When a replication fork encounters a damaged DNA (e.g., thymine dimer), it stalls. Replication may reinitiate further along leaving a gap opposite the damaged region. Such a gap can be lethal to DNA but it may provide the ssDNA for binding of RecA, and then draw the double-stranded region from the homologous DNA into a triple stranded structure where recombination can take place. For this, an undamaged strand from the other helix moves into the gapped region and it is replaced by the gap filling action of DNA polymerase (Fig. 2.17). This process can be repeated every time a replication fork moves past a damaged region of DNA without actually correcting it. Thus the damage would be

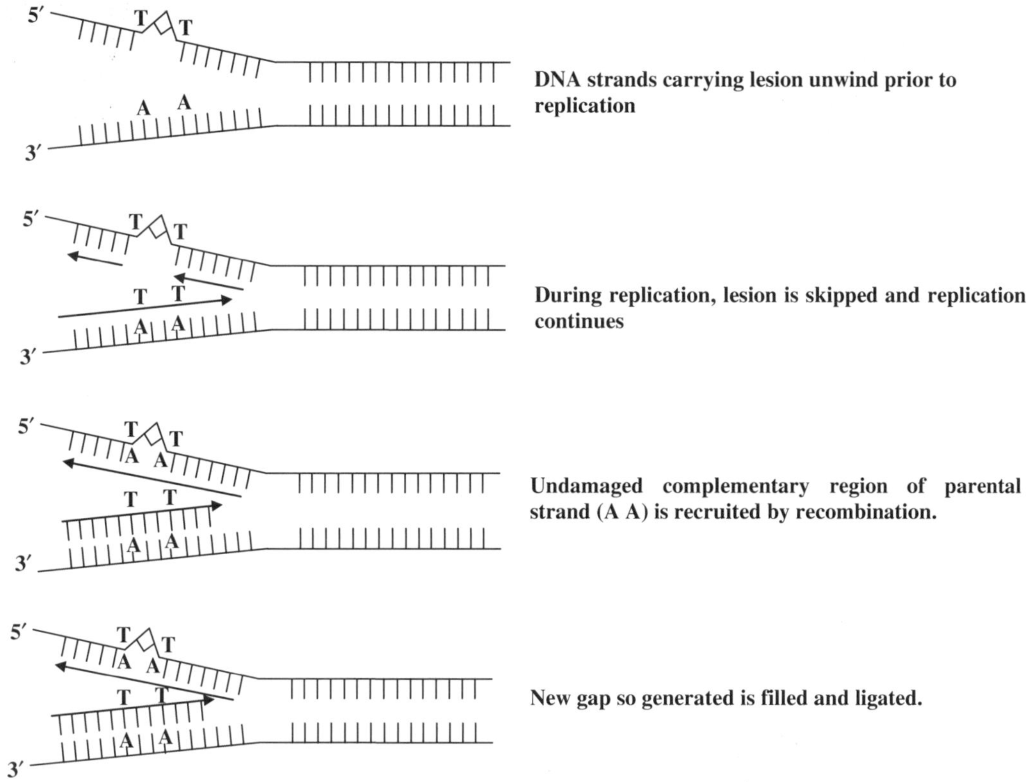

Fig. 2.17 Post-replication/recombinational repair

ultimately diluted out, as very few cells will carry the original damage.

In the leading strand, however, DNA polymerase will stop at the damage site, but DnaB helicase will remain associated and the lagging strand synthesis will continue. The gap so created can be handled by RecFOR proteins that help in loading the RecA to the single-stranded DNA, followed by the invasion of this filament in the double-stranded lagging strand. Reloading of DnaB and the replication apparatus is followed via PriC pathway. As a consequence, the thymine dimer is now located opposite good lagging strand and the gap is opposite a good leading strand (Fig. 2.18a). A perfect target for gap filling thus becomes available.

Another model has been proposed to explain this repair pathway, which begins the same way as described above. In a subsequent step, however, the two newly synthesized strands are proposed to pair with each other, a reaction promoted by RecFOR and RecA. This creates a branch that can retract backward (fork regression) to generate a type of Holliday junction that has been appropriately referred as "chicken foot". In this paired structure, the replicated lagging strand is somewhat longer than the blocked leading strand. The 3′end of the latter provides a primer for repair DNA polymerase that extends it until the two strands are of the same length. The chicken foot Holliday junction could then move forward possibly with the help of RecG or other helicases. With the establishment of bonafide replication fork on either side of the damage, replication could continue. This will require the reloading of DnaB helicase and other replication proteins, perhaps by PriA–PriB, DnaT–DnaC pathway (Fig. 2.18b).

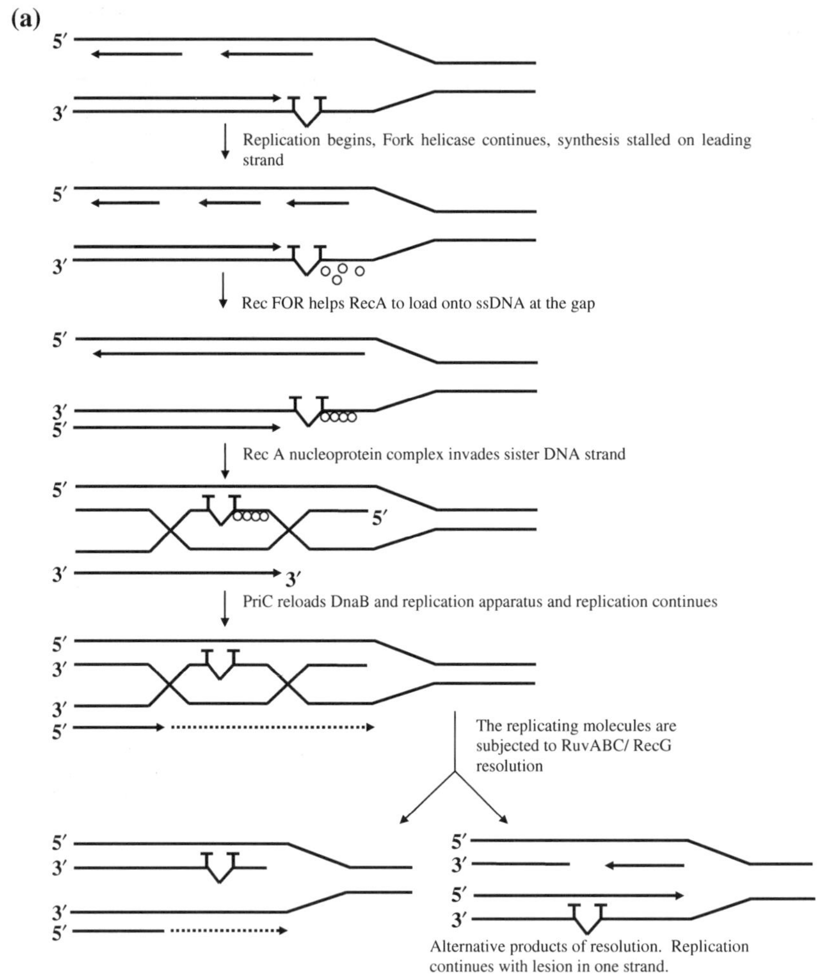

Fig. 2.18a Translesion repair involving leading strand and recombinational pathway

2.9.5 Inter-Strand Cross-link Repair

Many chemicals like light-activated psoralens, mitomycin, cisplatin, and EMS can cause inter-strand cross-links by promoting covalent bonding of two bases in the opposite strands. Such cross-links can neither be repaired by excision repair (a ds break will ensue which may be fatal) nor by recombination repair as movement of replication fork will be impeded by such cross-links. Thus, a combinatorial approach is employed.

First the UvrABC endonuclease nicks in one strand on either side of the cross-link and then the gap is widened by the 5′ exonuclease activity of DNA polymerase I. The gap so created is repaired by recombination repair by using one undamaged opposite strand. Now the DNA damage is confined to only one strand which can be repaired by nucleotide excision repair pathway.

2.10 Site-Directed Mutagenesis

With the advent of gene cloning technology, several genes have been cloned in appropriate vectors, sometimes even without the proper information on their function and regulation. A number of methods were subsequently

(b)

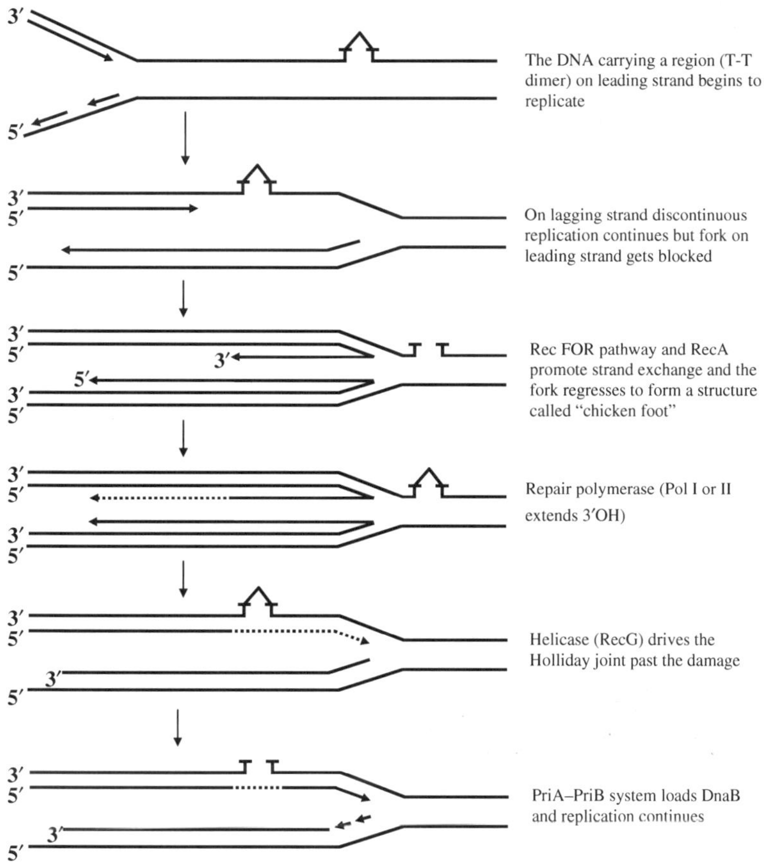

Fig. 2.18b Translation repair involving leading strand and recombinational pathway

developed to introduce mutations in such genes. Since these mutations are directed to a particular gene such methods are referred to as site-directed mutagenesis. Introduction of such mutations has facilitated a better understanding of a gene function and its regulation. These mutations can be broadly classified into three categories:

(1) Simple deletions or insertions.
(2) Systematic deletions, insertions, or substitutions of cluster of bases.
(3) Substitution of single bases.

2.10.1 Simple Deletions or Insertions

The removal of a sequence from cloned DNA is the simplest method to introduce changes in a defined sequence. Since it involves digestion with appropriate restriction enzymes, the restriction map of complete DNA sequence greatly facilitates the choice of enzyme/s. In the simplest form, it consists of removal of a stretch of DNA between restriction sites of an enzyme that occur only on cloned DNA and not on the vector DNA, and then by closing the digested plasmid DNA by ligation. If digestion generates complementary cohesive ends or blunt ends the ligation is direct and simple. If, however, incompatible ends are generated, Klenow fragment of Pol I, T_4 DNA polymerase, S1 nuclease, or mung-bean nuclease can be used to produce blunt-ended molecules that can be circularized by DNA ligase. By the same strategy a heterologous DNA sequence can be inserted at or

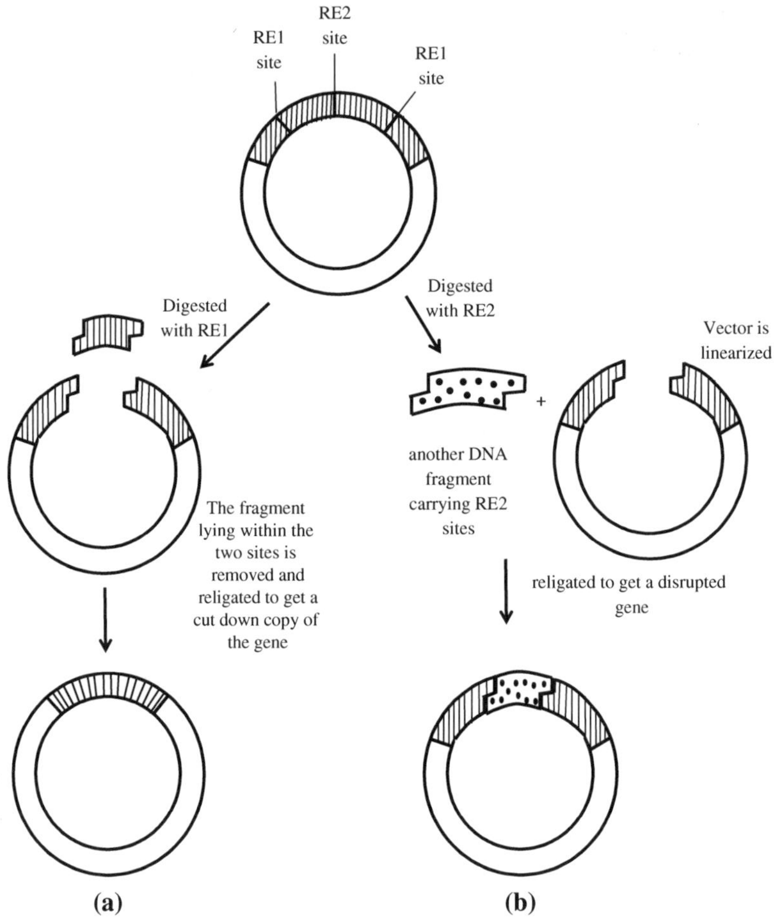

Fig. 2.19 Modification of a cloned DNA segment (gene ▥). (**a**) Deletion: wherein, the gene is digested with a RE(1) having two sites. The digested DNA is religated to get a cut-down copy of the gene. (**b**) Insertion: wherein, the gene is digested with a RE(2) having a single site and is ligated with another DNA fragment (often an antibiotic resistance cassette) generated by a similar enzyme which would also disrupt the gene

between the restriction sites in the cloned gene (Fig. 2.19).

The range of such mutations may, however, be very limited because of the non-availability of desired restriction sites and the undefined end points of deletions or insertions. Three other methods have been developed. These are:

1. Creation of unique restriction sites at defined locations by oligonucleotide-mediated mutagenesis and to use these sites for generating deletions and insertions.
2. Generation of systematic deletions and insertions in a region of interest and screening the mutants at desired locations by restriction mapping and/or DNA sequencing.
3. Oligonucleotide-mediated "loop out" mutagenesis for deletions and "loop in" mutagenesis for insertions.

Mutations so obtained are especially useful in identifying the upstream regulatory elements, such as size and location of the promoters and enhancers, etc. Such truncated cloned segments are also suited for DNA sequencing exercise.

2.10.2 Oligonucleotide-Mediated Mutagenesis

Unlike most other methods of mutagenesis, this can be used to generate mutation at a precise site. Thus, it can be applied for altering individual codons in a coding sequence or generate defined changes in a regulatory sequence or even in construction of new vectors and chimeric genes.

This essentially consists of: (1) cloning of a desired DNA fragment into phage M13 vector,

(2) preparation of single-stranded DNA from the recombinant phage, (3) designing and synthesis of mutagenic oligonucleotides, (4) hybridization of mutagenic oligonucleotide to the target DNA, (5) extension of hybridized region by DNA polymerase, (6) transfection of susceptible bacteria, (7) screening of plaques for those carrying desired mutation, (8) preparation of single-stranded DNA from the recombinant (mutagenized) phage, (9) confirmation of desired mutant by DNA sequencing, (10) recovery of mutated fragment from double-stranded replicative form, and (11) substitution of wild type DNA by mutagenized fragment (Fig. 2.20a). Each of these steps would require standardization before this method can be optimally utilized.

For example, the length of oligonucleotide (17–19 nt), its complementarity with the target DNA, and the position of the mismatched nucleotide (in the center or at one or two nucleotide positions immediately 3′ of the center) are all important considerations.

cIn a variation of this method, two primers, the phosphorylated mutagenic oligonucleotide and upstream universal sequencing primer (need not be phosphorylated) have been successfully employed.

In another method known as Kunkel method, the gene of interest is cloned in a phagemid vector and introduced into a *dut*$^-$, *ung*$^-$ *E. coli* strain. Also, the host cell is co-infected with a helper phage, so that the phagemid could be allowed to undergo replication. During this process, single-stranded template strands may bear a few uracil residues, as in such an *E. coli* strain, uracil cannot be removed from DNA. The mutagenic oligonucleotide is now annealed to single-stranded template and in vitro DNA synthesis is carried out to convert it into a dsDNA. Such a DNA will carry a heteroduplex region. When such a phagemid is transformed into an *ung*$^+$ strain, the template will undergo destruction due to the removal of uracil. Many of the resulting phagemids may carry the desired mutated strand (Fig. 2.20b).

When a whole variety of mutants need to be created, pools of degenerate single-stranded oligonucleotides can be used. This is applied when all the codons that are to be altered lie within a short stretch of contiguous nucleotides. An alternative method to generate pool of mutants is "cassette mutagenesis" in which wild type sequence is replaced by synthetic double-stranded oligonucleotides. Different types of cassette mutagenesis can be employed. In one method, two sets of oligonucleotides are synthesized, one consists of a sequence that corresponds to the wild-type target DNA strand, and the other comprises a degenerate pool, complementary to the other strand and carrying the desired mutation. These oligonucleotides are annealed (Fig. 2.21). In the second method, both the complementary strands consist of mixed-sequence oligonucleotides and carry mutations at different sites. In the third method, on the other hand, two degenerate pools of oligonucleotides are synthesized that are complementary to the same strand of the target DNA. The members of the two pools carry complementary palindromic sequence at their 3′ termini and when annealed form partial hybrids. Such hybrids are converted to blunt ended double-stranded DNA by Klenow fragment, forming tail-to-tail dimers. These are then digested to generate unit length cassettes. Once these cassettes are synthesized from any of these methods, they are inserted in place of target sequence and transformed into competent bacteria (see Fig. 2.21). While the cassettes produced by the first two methods may have to undergo mismatch repair, in the third, this does not happen, as the cassette is a perfect homoduplex.

2.10.3 In vitro Mutagenesis of Double-Stranded/Single-Stranded DNA with Chemical Mutagens

In order to carry out random localized mutagenesis, a short fragment of double-stranded DNA can also be treated with mutagens like nitrous acid and hydroxylamine. The population of mutagenized fragments is recloned in a vector that may carry the remainder part of the wild-type gene. The low frequency and narrow range

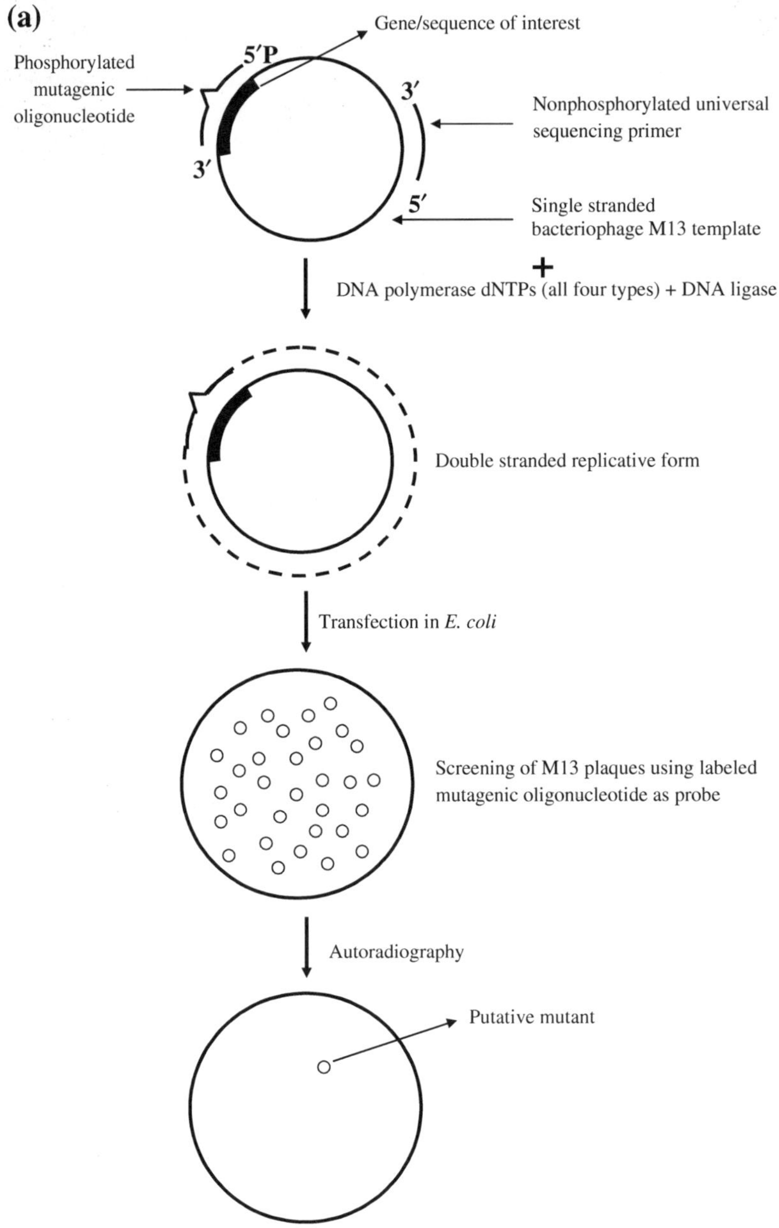

Fig. 2.20a Oligonucleotide-mediated site-directed mutagenesis

of mutations, however, limit the application of this technique.

In another method, circular double-stranded plasmid DNA is nicked at a random site with pancreatic DNase I and the nick is converted in a gap by exonuclease III digestion. Such a gapped, single-stranded DNA is exposed to sodium bisulfite (1.0–3.0 M) at a slightly acidic pH. After transformation of bacteria, replication of mutagenized DNA will have the replacement of original C–G base pair with T–A base pair, as sodium bisulphite will cause deamination of C to U. This method also leads to a very narrow range of mutations.

Fig. 2.20b Oligonucleotide-mediated site directed mutagenesis using Kunkel method

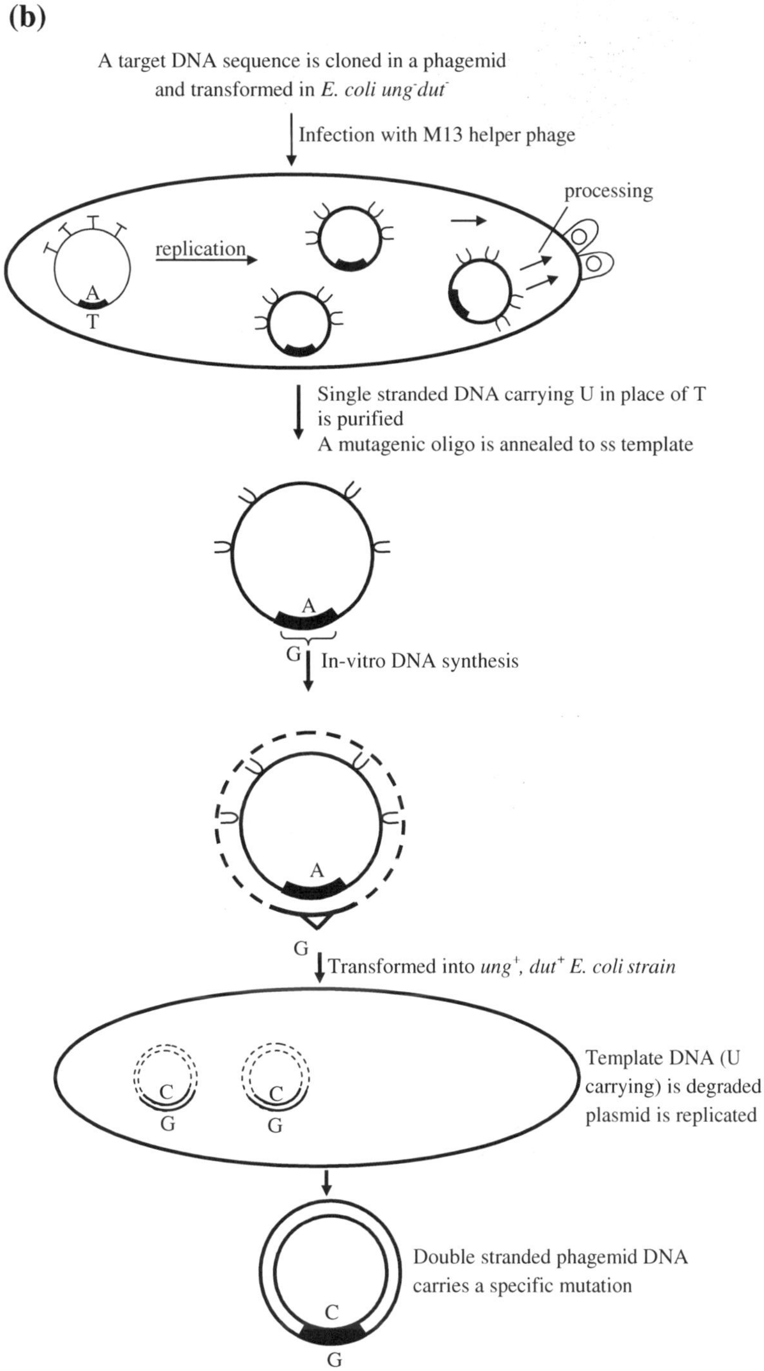

The single-stranded DNA in a phagemid can also be treated with chemicals like nitrous acid, formic acid, and hydrazine that can modify the bases without disrupting the phosphodiester backbone. After treatment, when the replication will ensue, the polymerase will tend to incorporate bases at random against the damaged base thus creating mutations. However, the frequency

Fig. 2.21 Cassette mutagenesis. The figure depicts one of the basic plans. For other alternatives refer text

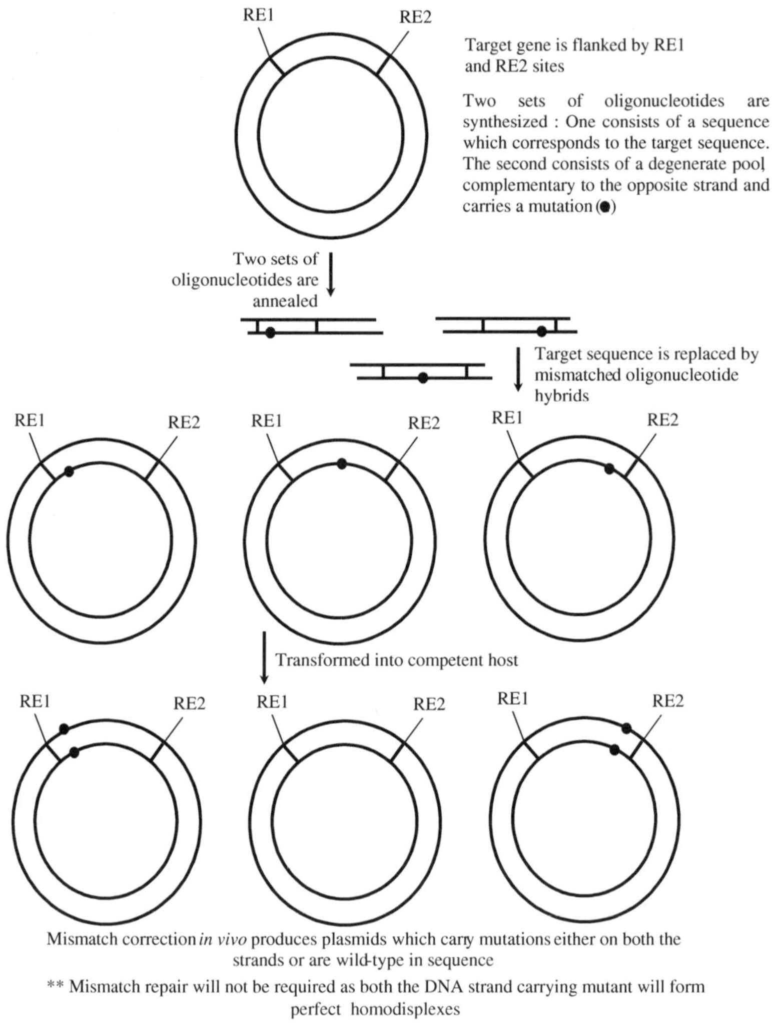

of useful mutations is not only low, even the possibility of multiple mutations largely restricts the use of this method.

Point mutations can also be introduced when gapped DNA are incubated in presence of DNA polymerase I and one of the four α-thiophosphate dNTPs. While these are effectively incorporated they are also not effectively removed by 3′–5′ editing function of Pol I. Thus, incorrect base can be incorporated at high frequency. The rest of the gap is then filled through second polymerization step in which all the four dNTPs are supplied. Another misincorporation method may employ AMV reverse transcriptase and base analogs. This is also based on the lack of 3′–5′ exonuclease function of the enzyme involved.

2.10.4 Systematic Deletion and Insertions

Methods are now available to generate small deletions or insertions in a more systematic way.

2.10.4.1 Linker Insertion Mutagenesis

Two methods of linker insertion mutagenesis are currently in use. Both require the generation of a set of circularly permuted linear molecules by

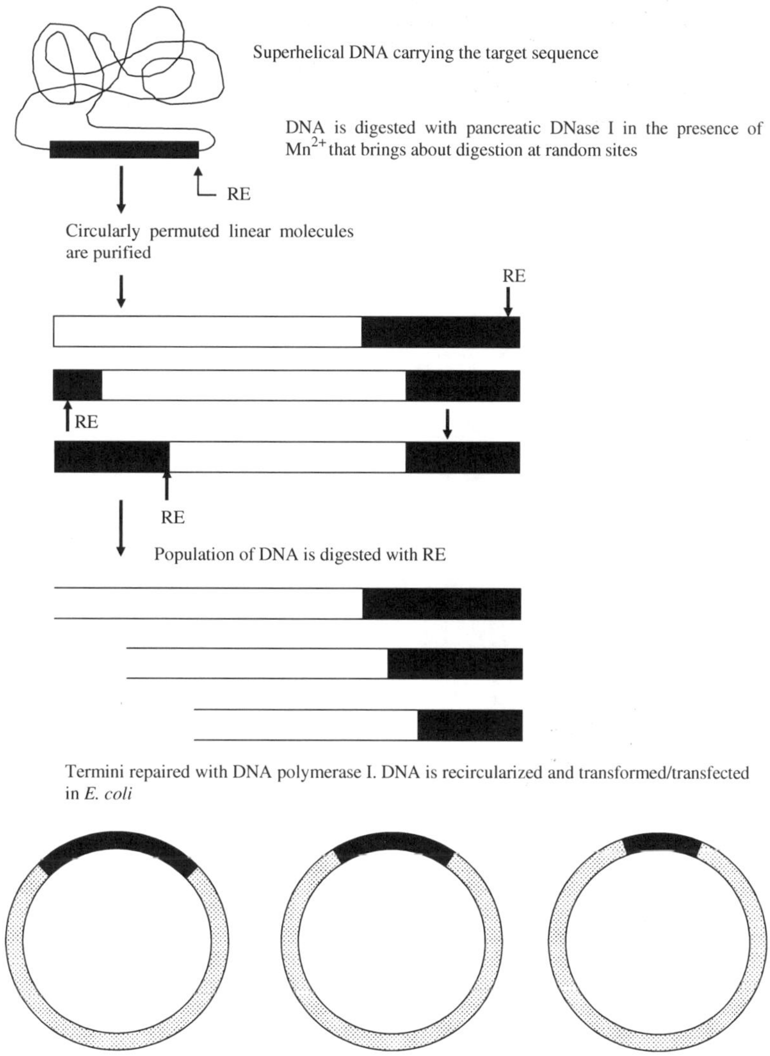

Fig. 2.22 Generation of nested deletions with pancreatic DNase I

random double-stranded cleavage of plasmid DNA.

In one method, this is carried out by partial digestion with pancreatic DNase I in presence of Mn^{2+} or Co^{2+} (Fig. 2.22). In the other, the partial cleavage with a restriction enzyme especially, those which create blunt ends serve the same purpose (Fig. 2.23). Both the protocols are fraught with problems that need to be taken care of, for example, if molecules with ragged termini or staggered ends are created they need to be first appropriately repaired.

Synthetic phosphorylated linkers carrying a restriction site not present in plasmid are then ligated directly to these blunt ends. Sometimes a linker carrying cohesive termini complementary to the one generated by a restriction enzyme can also be used. These inserted linkers can be converted to small deletions by cleaving with the appropriate restriction enzyme and then removing a small number of nucleotides from each linear end by Bal31 or a combination of exonuclease III and S1 nuclease or mung bean nuclease. By this method, synthetic linker is removed along with a few additional flanking nucleotides. Such linearized DNA is then circularized and transformed into an appropriate cell.

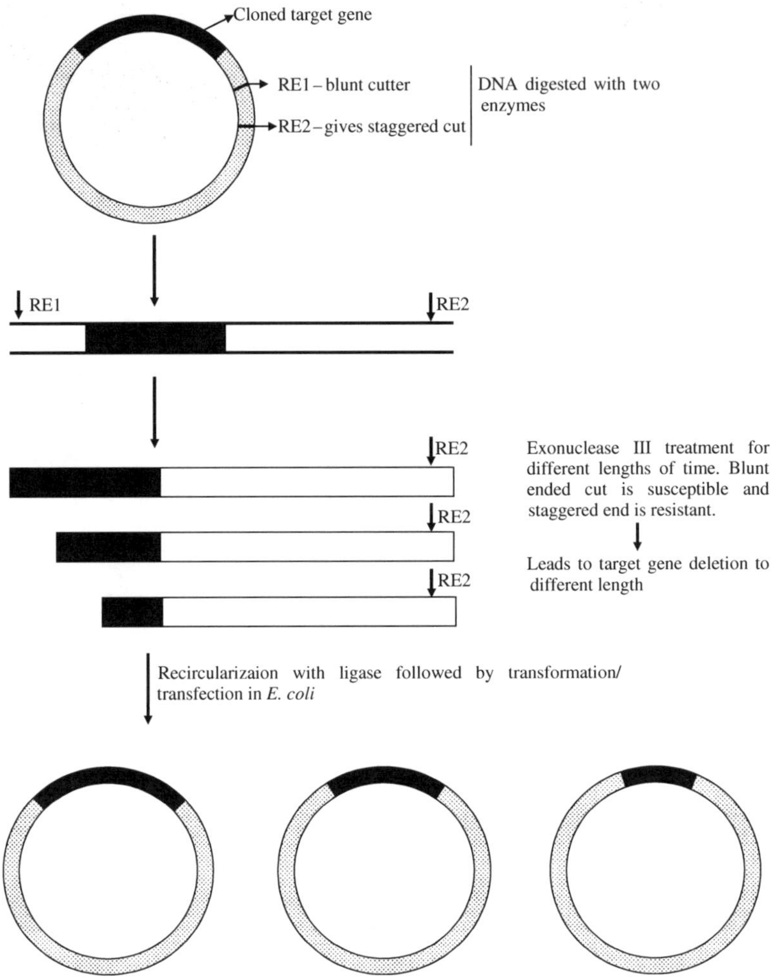

Fig. 2.23 Exonuclease III-mediated nested deletions in the target gene

Deletions can also be generated by exonuclease treatment. A circular DNA molecule is first digested by an endonuclease that cleaves only once. The linearized molecule is then shortened by limited digestion by an exonuclease, such as phage λ exonuclease. Such a DNA is circularized and transferred into a suitable bacterial cell. Insertions can also be introduced similarly. For example, digestion with *Hinf* I (G/ANTC) produces a 3 base long 5′ protruding end. The end is filled with Pol I and the molecule is circularized. This yields an insertion of three bases.

Such deletions are often very small, and can be enlarged by another procedure. In here, a linearized wild-type DNA is heteroduplexed with a deletion mutant, where a single-stranded loop will be formed on the former, at the site of deletion. The heteroduplex is then circularized and treated with S1 nuclease, which will specially cleave the single-stranded DNA in the loop region. The enzyme usually also removes 30 base pairs from the double-strand DNA, thus extending the original deletion.

Besides λ exonuclease, *E. coli* exonuclease III (Exo III) along with S1 nuclease can also be used. Since, nuclease Bal31 can act both as 3′ and 5′ exonuclease, this enzyme can replace S1 nuclease. The blunt ended molecules are circularized or if protruding ends are left they could be gap-filled with Klenow polymerase, circularized, and subsequently transformed.

The methods described above are not the only ones available to carry out site-directed mutagenesis, as newer methods are becoming available on regular basis. Interested readers shall be able to find them easily.

2.10.5 PCR Mutagenesis

Subsequent advances in molecular biology provided a precise method of mutating DNA by introducing specific changes in its nucleotide sequence. This consisted of the use of polymerase chain reaction or PCR. In here, a brief overview of this method is provided, as a lot of information is available in the literature for the interested readers. PCR mutagenesis allows us to modify and engineer any target DNA with ease and efficiency. This could lead to:

1. Introduction of insertion and deletion including the restructuring of the gene.
2. Altering one or a few nucleotides so as to change effectively the reading frame.
3. Randomly mutate a region of base sequence including a complete coding region, if desired.

For any mutagenesis reaction, DNA must be single-stranded so as to allow an oligo-nucleotide to bind to its complementary sequence in the target, the template. Another prerequisite is to clone the gene of interest. This is accomplished in a vector such as M13 or if a plasmid is used it can be denatured by alkali treatment, followed by neutralization step, or by heating. Most mutagenesis approaches incorporate a step that acts to eliminate the original template, leaving the mutated DNA to be transformed into the host cells. This facilitates the recovery of the clones derived from the mutated DNA.

2.11 Why are Mutations Important ?

As discussed earlier mutations are extremely valuable in any kind of genetic analysis. With advent of induced mutations and their rapid screening by employing microbial systems, their applications in understanding cell biology, physiology, gene function and its regulation, and molecular biology were also unraveled in a big way. We shall discuss some of these applications in detail:

2.11.1 Identification of Gene Function

Initially, and even today, the mutations are identified by a perceptible change in the phenotype. However, with the microorganisms, where either such markers are very few or are not easily distinguishable, biochemical techniques were employed to find out the differences. A large number of auxotrophic mutants, mutants with altered carbohydrate (or other nutrient) utilization, and those resistant to antimicrobial compounds were characterized this way. The use of radiolabeled substrate molecules led to the identification of transport-defective mutants. For example, *E. coli* mutants can be isolated that fail to grow on lactose. This could either be due to defect in the enzyme β-galactosidase or to the non-entry of lactose due to permease mutation. These two classes of mutants can be differentiated, say for example, by growing them on a medium containing labeled lactose. While the former will show the entry of lactose (label), the latter will fail to do so. Also, mutants resistant to analogs of some essential metabolites helped in isolating regulatory and/or permease mutants. Designing of a simple screening method for detecting a mutation still remains an important step.

Conditional mutants, such as temperature-sensitive mutations are especially useful in identifying essential genes, because mutations in these genes will not permit the survival of the cell. For example, gene functions required for DNA synthesis were identified by mutations with their temperature sensitive nature. Such mutants could grow like non-mutants at permissive temperature but behaved as mutants only at restrictive temperature. It should be clear that otherwise such mutants and therefore the underlying function of a gene could not have been revealed.

2.11.2 Elucidation of the Metabolic Pathways

Characterization of a set of mutants all belonging to one final product may even reveal its

Table 2.4 Identification of mutants by their growth behavior on different culture filtrates

Mutants type	Compound overproduced/released	Growth of the mutants in the culture filtrate
a	X	None
b	A	a
c	B	a, b
d	C	a, b, c

Note If compound D is added to the medium it supports the growth of all the mutants as well as the wild type, suggesting that it is the final product of this pathway

metabolic pathway. This is based on the fact that mutation by affecting a particular enzyme in a chain of reaction, will introduce a block in the conversion of a substrate (or intermediate) to next compound. This leads to the overproduction of the compound prior to the block, which may even be excreted out into the culture medium. The culture filtrate of this mutant, therefore, will allow the growth of another mutant in which the synthesis of the said compound is mutationally inhibited (Table 2.4). Analyses of different mutants will, therefore, lead to the identification of the steps in the metabolic pathways. Readers may recall the classical work of Beadle and Tatum in 1940s on *Neurospora crassa* that led to "one gene-one enzyme" hypothesis. A hypothetical pathway for the synthesis of compound 'D' has been depicted in Fig. 2.24.

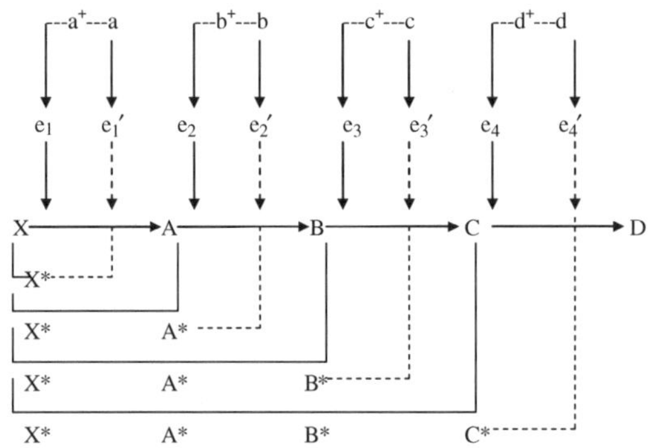

Fig. 2.24 The suggested pathway for the synthesis of final product D: The *upper part* explains the role of different genes and their mutants, which in turn through their encoded enzymes, control the operation of this metabolic pathway; the *lower part* represents the distribution of label in different compounds, if different mutants are fed on substrate X which was radioactively labeled

Subsequently, however, more refined techniques, like feeding on radiolabeled substrate and chasing the distribution of label in different compounds in different mutants, were employed for this purpose.

2.11.3 Understanding the Regulatory Pattern of a Metabolic Pathway

We have seen earlier that mutation can alter a protein and thereby a particular phenotype. However, mutations can also alter either the amount of a protein that is synthesized or the way its synthesis responds to an environmental signal. Such mutations can identify the regulatory systems. For example, the three genes of *lac* operon in *E. coli*, *lacZ*, *lacY*, and *lacA* are expressed at low (basal) levels, unless the cells are required to grow in the presence of lactose as the sole carbon source. When faced with lactose as carbon source, the enzymes coded by the three genes, viz., beta-galactosidase, permease, and transacetylase, are made to catabolize lactose, respectively. Such a system is known as inducible, logically so, as the presence of lactose induces the expression from the above described three genes. One could isolate mutants, in which these enzymes are always present (constitutive) irrespective of the presence or absence of lactose in the growth medium. This suggested that there is another gene (a regulator), which responds to the availability of lactose and accordingly switches on or switches off the concerned genes.

As the mutated regulatory gene is making the system constitutive, in a wild-type situation its product must be keeping the genes off. There are other sugar utilization systems known (*ara* operon, for example) where a mutation in the regulatory gene converts the system into a non-inducible one, thus suggesting that regulatory gene product is required to activate the genes. The former is thus called as negatively inducible and the latter positively inducible system.

2.11.4 Correlation of an In Vitro Biochemical Activity with an In Vivo Function

E. coli enzyme DNA polymerase I (Pol I) is capable of synthesizing DNA in vitro and it seemed likely that it carries out the same function in vivo. In an *E. coli* mutant (*polA*), where the activity of Pol I was reduced to 50 fold, no effect either on growth rate of the cells or their ability to synthesize DNA in vivo, was seen. This suggested that Pol I cannot be the only enzyme required for DNA synthesis. From the cell extracts of *polA* mutants, activity of two more DNA polymerases, Pol II and Pol III was detected. Later a temperature-sensitive *dnaE* (=*polC*) mutant was isolated in *E. coli*, which could synthesize DNA at the permissive temperature of 30 °C but failed to do so at the restrictive temperature of 42 °C. Enzyme analysis of this mutant revealed active Pol I, Pol II, and Pol III at 30 °C but at 42 °C only Pol III was completely inactive. This clearly indicated that Pol III is the in vivo DNA synthesizing polymerase.

2.11.5 Identification of the Site of Action of an External Agent

Antibiotics are known to affect different intracellular targets. Rifampicin, for example inhibits RNA synthesis. However, from this effect it cannot be ascertained whether the inhibition is due to its binding to DNA or to the RNA polymerase. Mutants resistant to rifampicin (Rif^r) were isolated and characterized. In most of these mutants RNA polymerase was slightly altered which suggested that rifampicin acts by binding to RNA polymerase.

2.11.6 Indication of a Relationship Between Two Systems

In microbial world, lots of studies have been carried out on host–parasite interactions, such as bacteria–bacteriophage systems. Often host mutants have been isolated on which a phage fails to grow. For example, on some *E. coli* mutants unable to metabolize malodextrins (*lamB*), phage λ fails to grow. This inability is not seen on mutants unable to metabolize other sugars. Also, other phages adsorb normally on to (*lamB*) mutant, clearly suggesting that *lamB* gene product is specifically required for λ adsorption.

2.11.7 Revealation of Protein–Protein Interaction

Very often a particular phenotype is controlled by more than one gene. For example, a particular process may be controlled by two genes *a*, and *b*, contributing the proteins A and B, respectively. In order to complete the process, either the two proteins may act in a step-wise manner or the two may interact to produce a single functional unit. The latter can be revealed if revertants are sought in the mutations of the two genes. For example, revertants of *a* will mostly belong to the gene *a*, but some may belong to gene *b*, and similarly those of *b* will be generally confined to *b* but some may be traced to gene *a*. Such a situation explains the interaction between two proteins, A and B.

2.11.8 Estimation of the Number of Gene Controlling a Phenotype and the Functional Differentiation Within a Gene

When large number of mutations are isolated and characterized, it is often observed that more than one gene can affect the same phenotype. These mutations can be classified into non-allelic or allelic types by a simple complementation test as discussed earlier.

The non-allelic mutations can be easily differentiated because in a trans-complementation test, they will always give a non-mutant phenotype. Thus, any time an independent mutation affecting a known phenotype is isolated it has to be put through a complementation test so as to be classified accordingly.

However, for mutations, which belong to the same gene, as in some situation, a gene may have an internal differentiation into functional units or cistrons. A complementation test between two members of the different cistrons will be positive but if they belong to the same cistron they will fail to complement. Such an intragenic complementation combined with rare intragenic recombination has led to fine structure maps, wherein the mutational site (muton) and/or recombinational sites (recon) have been mapped. This has allowed the better functional and structural understanding of a gene.

2.11.9 Basic Source of Variability

As we have discussed above the process and mechanism of mutation, it must have become amply clear that it constitutes the basis of genetic variability. Thus all allelic variations are contributed by mutations which are further expanded by gene recombination. These two, therefore, have played a critical role in biological evolution.

2.12 Reversion and Suppression

As discussed earlier, a number of mutagens work in reversible order. Thus a change brought in one direction, for example, GC → AT can be reversed by AT → GC change. Such a change can restore the original base sequence, and, therefore, the gene function and is called as reverse mutation. Very often the change brought about by a mutation is restored by a second mutation, taking place elsewhere in the DNA. The basis of such a mutation, referred to as suppressor mutation, is to relieve the effect of first mutation. The reversion frequency is the fraction of cells in a population of mutants that regain original phenotype per generation.

The chances of an exact reversion (reversing the earlier mutation event) are very low because the mutagen will have to act exactly on the changed base, something that normally does not happen. The generalized reversion frequency, as also the mutation frequency, is $\sim 10^{-7}$. But if a gene has 200 sites which can give rise to mutation, the probability of such an event will be $1/200 \times 10^7$ or $½ \times 10^9$. The reversion frequency for any particular mutation may vary widely but it is often high enough to be expected from a return of wild-type base sequence. This gave rise to the idea that reversion may occur due to second site or suppressor mutation. The latter may be differentiated into intragenic or an intergenic suppressor mutation.

2.12.1 Intragenic Suppressor

A good deal of biochemical data exists to demonstrate that the effect of a mutation can be reversed by introducing another mutation. This may involve compensating for an earlier change, e.g., an addition can be reversed by deletion and vice versa, except that the second change should not take place very far from the first site and the segment in which these two changes occur must not lead to any substantial alteration in the protein product. Alternatively, this may involve compensation of charge between the participating amino acids or for other interactions. Reversion is a useful test to determine whether the phenotype is due to a single site mutation, because if the phenotype is controlled by more than one gene the reversion will become exceedingly rare and may not be seen at all. For example, if an auxotrophic mutation happens to show growth impairment even when its requirement is fulfilled, this could be explained either due to one mutation having multiple effects or two independent mutations. This situation can be resolved by seeking reversion for prototrophy. If the revertant grows at normal rate, the two effects were due to same mutation but if it continues to grow slowly there were two mutations from the beginning.

Reversion has also been employed to test for possible mutagenic and/or carcinogenic activity of a compound. Readers must have been familiar with one such test called as Ames Test.

2.12.2 Intergenic Suppressor

Such mutations constitute another class which can negate the effect of an earlier mutation. They function through a variety of mechanisms.

In one of the common pathways, the cell's translational machinery is modified in such a way that even a mutant codon is read (misread) leading to the synthesis of a protein product from a mutant gene. Many such suppressor mutations called informational suppressor involve tRNA that not only suppresses the original mutation but many others which are suppressor-sensitive mutations. The latter is a type of a conditional mutation in which the mutant shows a wild-type phenotype if suppressor is present, but if not, then the mutant phenotype is expressed. Such strains producing suppressor are designated as *sup* and those not producing them as *sup*^{o}. Suppressor-sensitive mutations are of two types: nonsense or chain termination mutations and missense or amino acid substitution mutation. A rare class of frame-shift suppressors has also been detected. There are apparently strong context effects on the ability of a tRNA anticodon to correctly translate a codon. For example, it is possible that the normal tRNA-UUG translates a UGA codon, provided the codon is followed by another codon beginning A. The ability to ignore context effects in some measure depends on the correct modification of the base just before the anticodon (i.e., the 3′ base). When it is unmodified, suppressor tRNA molecules function much less efficiently. Some informational suppressors may be traced to ribosomal genes that may cause the incorrect translation of mRNA at a high frequency.

Very often a phenotype results from protein–protein interaction and a mutation in one gene may simply disrupt this interaction. In such a mutant, mutation selected in the second gene may restore the interaction. Such suppressors also known as interaction suppressors are usually allele-specific and function only with a defined subset of mutations. These have been employed to define sites on the two proteins that interact, as we discussed earlier.

Many mutants produce protein product with lower activity and can be suppressed by overproducing the mutant protein. For example, a *lacZ* missence mutation can be suppressed by promoter-up mutations. One way of increasing the protein product can be to check its degradation.

Another type, called bypass suppressors in fact, activates a new pathway and thus eliminates the need for the mutant gene. The *sbc* mutation in *E. coli* reverses the effect of *recBCD* mutations, and *ebg* mutation codes for a new enzyme that can replace defective β-galactosidase in *lacZ* strains. Such mutants may furnish an important component the synthesis/activity of which is blocked in another gene. Some of them can even turn on an inactive ("cryptic") gene that has a similar function.

Some mutations can be suppressed by general changes in the cell physiology. For example, many mutations may produce functional but unstable products and the same may be suppressed by mutations that increase the cellular concentration of molecules that could stabilize such proteins.

All such intergenic suppressor mutations, therefore, can mask the presence of a mutant gene. The presence of an intragenic or an intergenic suppressor can be easily demonstrated by crossing the strains harboring such suppressor with the wild-type strain. The product of such a cross will show at least 25 % mutants if there was an intergenic suppressor and no mutants if there was an intragenic suppressor present.

2.13 Directed Mutation

Both the fluctuation test and replica plating conclusively proved that mutations are random or non-directed. Such a notion prevailed for a very long time until Cairns and others demonstrated that under certain specific circumstances bacterial mutations do appear directed. For example, they found that Lac^- strain of *E. coli* reverted to Lac^+ at high frequency when grown in the presence of lactose as compared to when exposed to glucose as the sole carbon source. If the two conditions are compared, reversion under the former is advantageous for cell's survival as compared to the latter. Thus, such a mutation can be classified as directed or adaptive mutation. This observation led to some debate on the resurgence of Lamarckian view of evolution but available data suggest that it may not entirely be so. Such mutations seem to happen when a population of cells fails to grow (stationary phase) rather than when they are actually killed. It is clear that the absence of net growth does not necessarily mean the absence of all metabolic activities and the latter may include DNA replication connected with DNA repair. Evidence are mounting that under such a state, cells are more susceptible to mutation and obviously to those which provide a selective advantage (for example $Lac^- \rightarrow Lac^+$ in presence of lactose as the sole carbon source).

Though a clear mechanism(s) of stationary-phase mutagenesis has not emerged, genetic evidences suggest that there are at least two principal pathways operating in the cells when they are exposed to growth limiting environment. In majority of the cases, there could be a compensatory -1 frame shift mutation. This requires the recombination proteins, such as, RecA, RecBC, RuvA, RuvB, and RuvC, and induction of the SOS response regulon, particularly error-prone DNA polymerase IV (Din B).

The other pathway implicates amplification of the involved locus (e.g., *lac*) into a tandem array of 20–100 copies leading to transient hypermutation. Adaptive amplification requires RpoS, the transcriptional activator of $\sim$50 stationary phase/starvation- and general stress-responsive genes.

Mutator Gene

Although mutations are important for all living cells, a strong genetic control has evolved to keep the mutation frequency low. This will ensure higher stability of a gene. The *mut* genes, which have been assigned this function, can increase the mutation frequency of the other genes when they themselves are defective.

The mutation, *mutD*, for example, in the epsilon subunit of Pol III, affect the proof-reading function ($3' \rightarrow 5'$ exonuclease) of DNA polymerase III (*dnaQ*). About 95 % of such mutations were of the GC → TA or AT → TA type. Similarly, *dam* mutation directing the mismatch repair system to the correct strand can lead to no or little methylation of the parent strand so that in some cases, correct base is excised and a new base is inserted to pair with incorrect base. Other *mut* genes, such as, *mutH*, *mutL* and *mutS*, also involved in methyl-directed mismatch repair have been described earlier under "DNA Damage and Repair".

An alternative class of mutation having the opposite effect, that is antimutator, has also been reported. They alter the replication machinery in such a way that the error rate is reduced below the normal 10^{-10}/base incorporated.

Further Readings

Benedicte M (2005) After 30 years of study, the bacterial SOS response still surprises us. PLoS Biol 3(7):e255

Carter P (1986) Site-directed mutagenesis. Biochem J 237:1–7

Griffith AJF, Wessler SR, Lewontin RC, Carrol SB (2008) Introduction to Genetic Analysis 9th Edition. W.H. Freeman and Company, N.Y

Hastings PJ, Slack A, Petrosino JF, Rosenberg SM (2004) Adaptive amplification and point mutation are independent mechanisms: evidence for various stress-inducible mutation mechanisms. PLoS Biol 2(12):e399

Janion C (2008) Inducible SOS response system of DNA repair in *Escherichia coli*. Int J Biol Sci 4(6):338–344

Klug WS, Cummings MR (2003) Concepts of Genetics. 7th Edition. Pearson Education Inc, San Francisco

Krebs JE, Goldstein ES, Kilpatrick ST (2006) Lewin's essential genes, 2nd edn. Jones & Bartlettl, Boston

Miesfield R (2001) DNA libraries and mutagenesis. www.biochem.arizona.edu./classes/bioc471a/571a

Morimatsu K, Kowalczykowski SC (2003) RecFOR proteins load RecA protein onto gapped DNA to accelerate DNA strand exchange: a universal step of recombinational repair. Mol Cell 11(5):1337–1347

Patel M, Jiang Q, Woodgate R, Cox MM, Goodman MF (2010) A new model for SOS-induced mutagenesis: how RecA protein activates DNA polymerase V. Crit Rev Biochem Mol Biol 45(3):171–184

Rice, K.P. and Cox, M.M. 2001. Recombinational DNA repair in bacteria: post-replication. Encyclopaedia of Life Sciences. Nature Publ. group

Roth JR, Kugelberg E, Reams AB, Kofoid E, Andersson DI (2006) Origin of mutations under selection : the adaptive mutation controversy. Annu Rev Micrbiol 60:477–501

Smith KC (2004) Recombinational DNA repair: the ignored repair system. BioEssays 26:1322–1326

Wright BC (2000) A biochemical mechanism for non-random mutations and evolution. J Bact 182(11): 2993–3001

3 Conjugation

From the time bacterial cells could be observed through a microscope, bacteriologists have tried to view and depict the cells indulging in sexual reproduction on a purely morphological basis. Although many of these studies suggested cell–cell contact the results could not be validated as they lacked a genetic basis. Designed crosses were conducted in bacteria by Sherman and Wing (1937) and Gowen and Lincoln (1942), but the characteristics taken into consideration were too complicated to evaluate their outcome. One of the hallmarks of sexual reproduction established in eukaryotes is the exchange of genetic material from two potentially different genomes. This process known as recombination produces new combinations of genes, allowing a mechanism of gene reassortment within the population and maintenance of genetic heterogeneity.

3.1 The Historical Cross

In the 1940s, Joshua Lederberg and Edward L Tatum selected the bacterium *Escherichia coli* K-12 to carry out a systematic study on genetic recombination. This bacterium was by now popular as a standard microbiological strain as it could grow on a simple nutrient medium (minimal medium). At the same time *E. coli* was amenable to induced mutagenesis and Tatum had already collected a fairly large stock of biochemical mutants including auxotrophs (single and multiple), phage/antibiotic resistant, and non-utilization of a particular carbon-source. When they set out to conduct their experiment they were clear that a rare recombinant, if formed, will have to be selected for and should be distinguishable from the spontaneous mutation.

In order to take care of the last two points, multiple auxotrophy in *E. coli* turned out to be very handy. First, they appeared to be stable, though each one of them is revertible to wild type independently, a simultaneous reversion of two or more markers will be possible only at an improbable frequency. For example, if one mutation reverts at a frequency of $\sim 10^{-6}$, the two simultaneous reversions will only occur at $\sim 10^{-6}$ x $\sim 10^{-6}$ or $\sim 10^{-12}$ frequency, scoring of which will become highly unlikely. Second, crossing two multiple auxotrophic strains should yield an all prototrophic recombinant class that should be selectable on the minimal medium.

They went about by setting up a cross between the strains 58-161 (requiring methionine and biotin: *met*, *bio*) with W677 (requiring threonine, leucine, and thiamine: *thr*, *leu*, *thi*). The results of this cross were published in 1946, where they demonstrated that if the growing populations of two strains are independently plated on minimal agar no colonies appeared (in other words, no reversion), but when the same number of cells are mixed and then plated on minimal medium, a few colonies appeared at a frequency of 10^{-6}–10^{-7} relative to the original cell number (Fig. 3.1). These colonies continued to grow on minimal

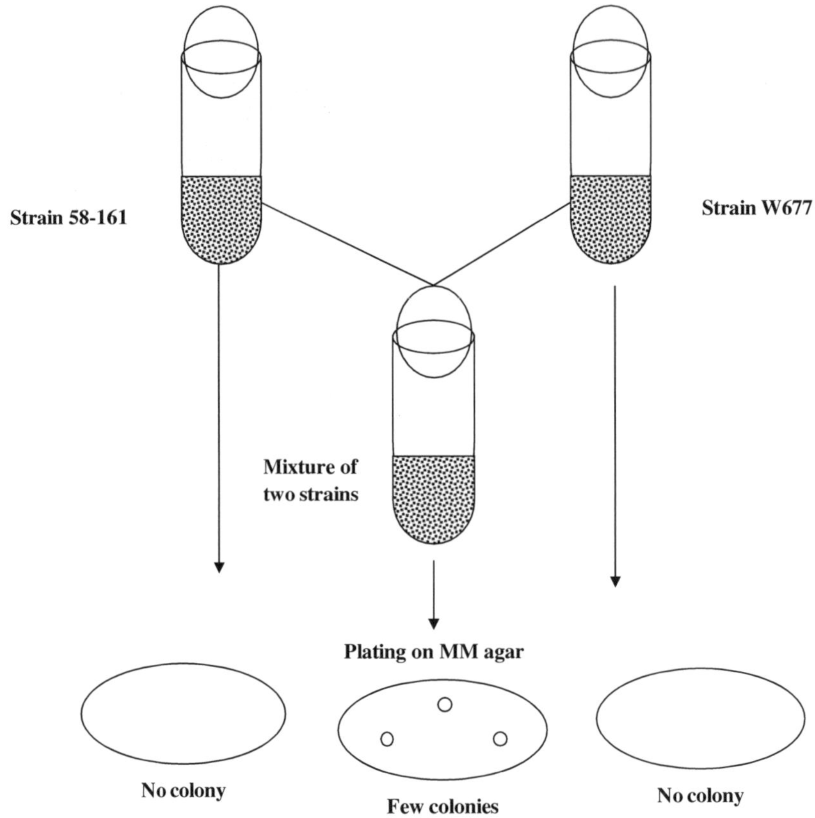

Fig. 3.1 Lederberg and Tatum's experiment to demonstrate conjugation in *E. coli*. Plating of suspension from a mixture of two strains produced a few colonies on MM agar. When two strains were plated separately no growth was seen, suggesting no reversion

agar on further transfers showing their stable nature. Lederberg and Tatum inferred that the two strains could have exchanged their genetic material and produce one possible class of prototrophic recombinants as shown below. Although the frequency of recombination was low it was easily distinguishable from any suspected reversion. The possible step underlying this event is depicted below:

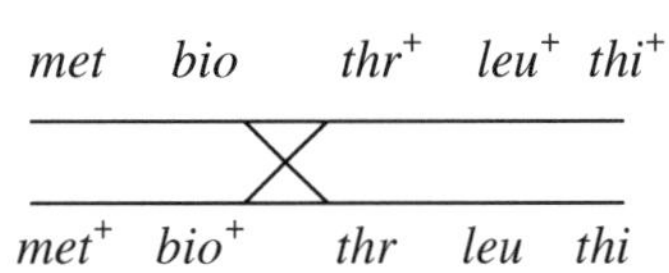

Though only one class of recombinant is observed (the one that is selected for, say on minimal medium), its complementary all auxotrophic class must also have been generated, as also the recombinants that will form due to crossing over between other genes. None of these classes would survive on minimal medium (selection medium). Thus the selection for a particular class of the progeny is an important consideration in the analysis of a bacterial cross. As the process involved the bringing together of single-celled organisms it came to be referred as conjugation. Thus in 1946, bacterial sexual reproduction was discovered through the demonstration of genetic recombination in *E. coli*.

That there is indeed a genetic exchange in the generation of these prototrophic recombinants was further established by the analysis of unselected markers. In such crosses the auxotrophic markers behaved as selected markers as the initial selection of recombinants is dependent upon

Table 3.1 Distribution of unselected marker ton^r/ton^s amongst the prototrophic recombinants from the cross 58-161X W677

Cross	Percentage distribution	
	ton^r	ton^s
thr^+ leu^+ thi^+ *met bio* ton^r × *thr leu thi* met^+ bio^+ ton^s	75	25
thr^+ leu^+ thi^+ *met bio* ton^s × *thr leu thi* met^+ bio^+ ton^r	36	64

their presence. The two strains used also differed in two other markers: ton^r/ton^s (resistance/sensitivity to phage T1), and ability/inability to utilize some β-galactosides (lactose): lac^+/lac^-. These characteristics are not important for the initial selection of the prototrophic recombinants and thus are referred as unselected markers. However, distribution of unselected markers can be studied in selected recombinants by subsequently either exposing them to (as in this example) phage T_1 or on an indicator agar. Lederberg and Tatum showed that distribution of unselected markers is clearly non-random, making it a fit case of genetic linkage (Table 3.1).

In the reciprocal cross depicted above, the unselected marker ton^r appears to have a tendency to go with thr^+ leu^+ thi^+, which in a normal genetic jargon would refer to a phenomenon that identifies linkage.

In 1950, Davis confirmed that the process requires cell-to-cell contact and thus can indeed be described as conjugation. His experimental setup consisted of a U-tube, the two arms of which can be separated by the placement of a sintered glass filter. He took the same two strains in the two arms of the U-tube wherein the free movement of cells was blocked due to the filter whose pore size was small enough to stop the passage of cells across it. A small amount of material was taken from the two arms periodically and plated on selection agar. No recombinants, however, were observed. When the filter was removed, so that the cells were free to move around and mix (could also be facilitated by applying pressure or suction), recombinants could be selected on minimal medium, as before (Fig. 3.2). This experiment clearly proved that genetic transfer requires contact between the two types of cells.

3.2 Compatibility

When studies on bacterial conjugation were extended several other aspects of this process came to light. In 1952, Joshua Lederberg, Cavalli and Esther Lederberg happened to use a derivative of strain 58–161 (*met*, *bio*) that failed to produce recombinants with the standard W677 (*thr*, *leu*, *thi*), suggesting thereby that strain 58–161 has lost its fertility. They proposed that though bacteria may not have an apparent morphological differentiation of sex the compatibility or mating type differences do exist. Thus one strain was referred as fertile or F^+ and the other F^- or sterile. For any cross to be successful, there has to be a participation of one F^+ and one F^- strain. While two infertile ones never mate the $F^+ \times F^+$ cross may give rise to the recombinants, albeit at a frequency still lower than $F^+ \times F^-$ frequency.

Hayes (1952) conducted crucial experiments to assign functional differences between the two cell types. He carried out two sets of $F^+ \times F^-$ crosses. In one, F^+ strain was treated with either UV-irradiation or the antibiotic streptomycin, whereas in the other the F^- strain was subjected to the same treatments. While in the first case the outcome of the crosses did not differ, in fact recombinant formation was even stimulated slightly by UV treatment, in the second, no viable recombinants were observed. This led Hayes to conclude that clear functional differences exist between the two cell types. While F^+ behaves as donor or male, the F^- functions as recipient or female. From the differential effects of these agents it could also be inferred that the transfer of genetic material is unidirectional, i.e., from donor to the recipient and the viability of

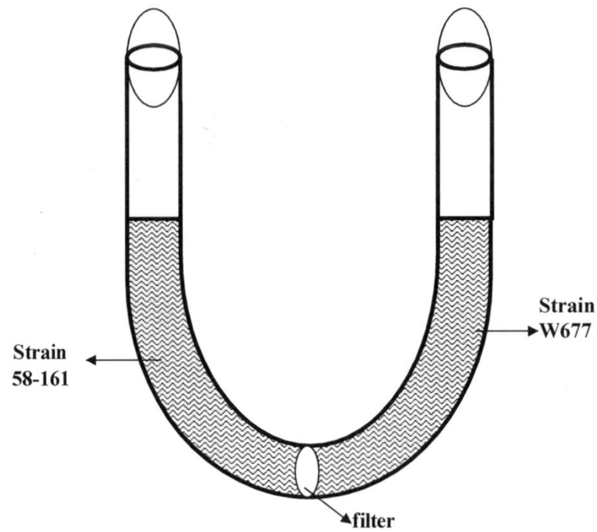

Fig. 3.2 Davis U-tube experiment to demonstrate that cell-to-cell contact is essential for gene transfer during conjugation. As long as the two strains remain separated by a filter no wild-type recombinants were obtained. Once this filter is removed so that mixing of the two populations is restored, recombinants can be selected on minimal medium as selection agar, as shown earlier

Table 3.2 Hayes experiment to demonstrate the functional differences between donor and recipient cells and the unidirectional flow of genetic information, i.e., from donor to recipient

S. No.	Matings		Recombinants
	Donor	Recipient	
1	Untreated control	Untreated control	Yes
2	Treated with streptomycin or irradiated with UV	Untreated Untreated	Yes Yes
3	Untreated	Treated with streptomycin	No
4	Untreated	Irradiated with UV	No

the recipient is important as this cell is going to process the transferred male genetic material and produce the progeny (Table 3.2).

Another important discovery made by Lederberg's group in 1952, was based on compatibility testing of the progeny of an $F^+ \times F^-$ cross. They showed that all the progeny were F^+, as if the fertility character is being infectiously transferred, independent of any other genetic marker. Sexual conversion of the progeny in the cross $F^+ \times F^-$ occurred at a frequency that is 10^6-fold higher than recombination (Fig. 3.3). This may create an impression that all F^- strains of *E. coli* will be converted into F^+ one day. In an independent experiment, however, it was also shown that when F^+ cells are grown to a high cell density, some cells may start behaving like F^-, with no concomitant change in the genotype. In other words, fertility function can be lost but the genetic markers remain unchanged. In classical genetics, such a situation is referred to "phenocopy" formation.

Hayes (1953) explained the ability of F^+ strain to transfer mating type characteristic efficiently and genetic marker less efficiently on the basis of F^+ strain possessing a sex or fertility factor or F factor. The rapid transfer classifies it as an extra-chromosomal element. Another feature conferred by this factor is that it prevents the cell to act as recipient.

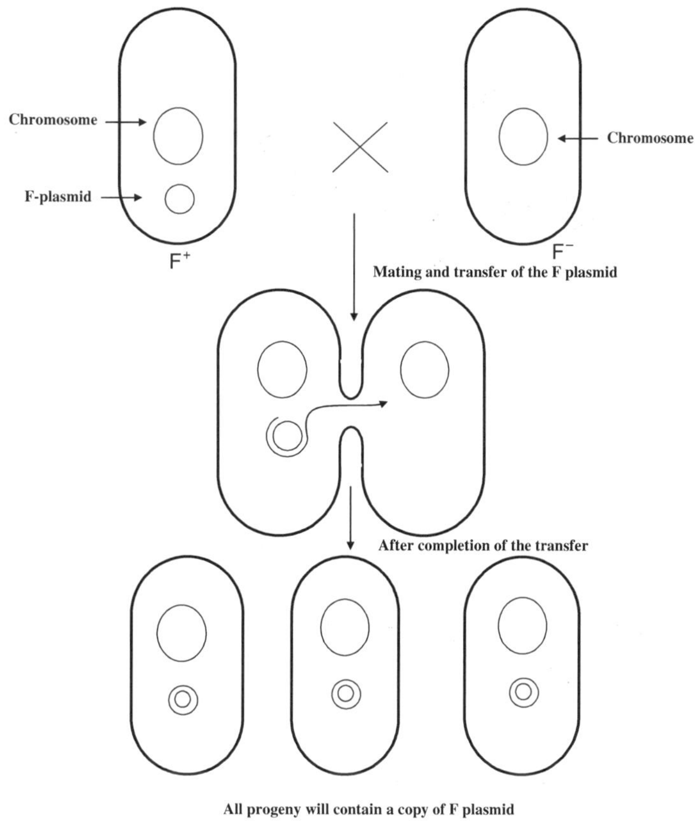

Fig. 3.3 A cross between F$^+$ donor and F$^-$ recipient leads to plasmid transfer and all the progeny will contain a copy of F plasmid (conversion to donor type)

3.3 Formation of Recombinants

The initial surge of activity in this field helped bringing gene inheritance in bacteria at par with already established phenomenon in eukaryotes. Important differences, however, also kept cropping up between the two systems. Analysis of a conjugational mating, for example, is very different from conventional genetic analysis. First, only those recombinants can be analyzed that are selected for, and second the immediate product of conjugation is not analyzed. What is examined are the colonies derived from the product formed due to transfer of the chromosome from the donor to the recipient. In bacterial crosses, the conventional diploid structure or zygote is never obtained. Lederberg's group had shown that if one or both the parents carry a *het* allele, one does find colonies, which behave as heterogenotes. Such heterogenotes are highly unstable producing stable haploid genotypes.

Hayes (1953) analyzed the prototrophic recombinants of the cross 58–161 × W677 and found that most possess the unselected markers of the strain W677, the earlier identified F$^-$ strain. Thus, there appeared to be a bias toward the markers of the recipient, except for those donor markers that are selected for. Although Lederberg's group had proposed that the entire male chromosome is first transferred, what is selected is utilized and the rest of it is lost. Hayes, on the other hand, discarded this suggestion based on its wastefulness. He in fact suggested that donor bacterium transfers its chromosome only partially and what is transferred only can be selected for. Also, no complete diploid zygote is formed instead the structure at best can be referred as merozygote/merodiploid/heterogenote (Fig. 3.4).

The analysis of bacterial reciprocal crosses is also different from that of eukaryotes. For example,

F$^+$ lac$^+$ a$^+$ b$^-$ X F$^-$ a$^-$ b$^+$ lac$^-$.

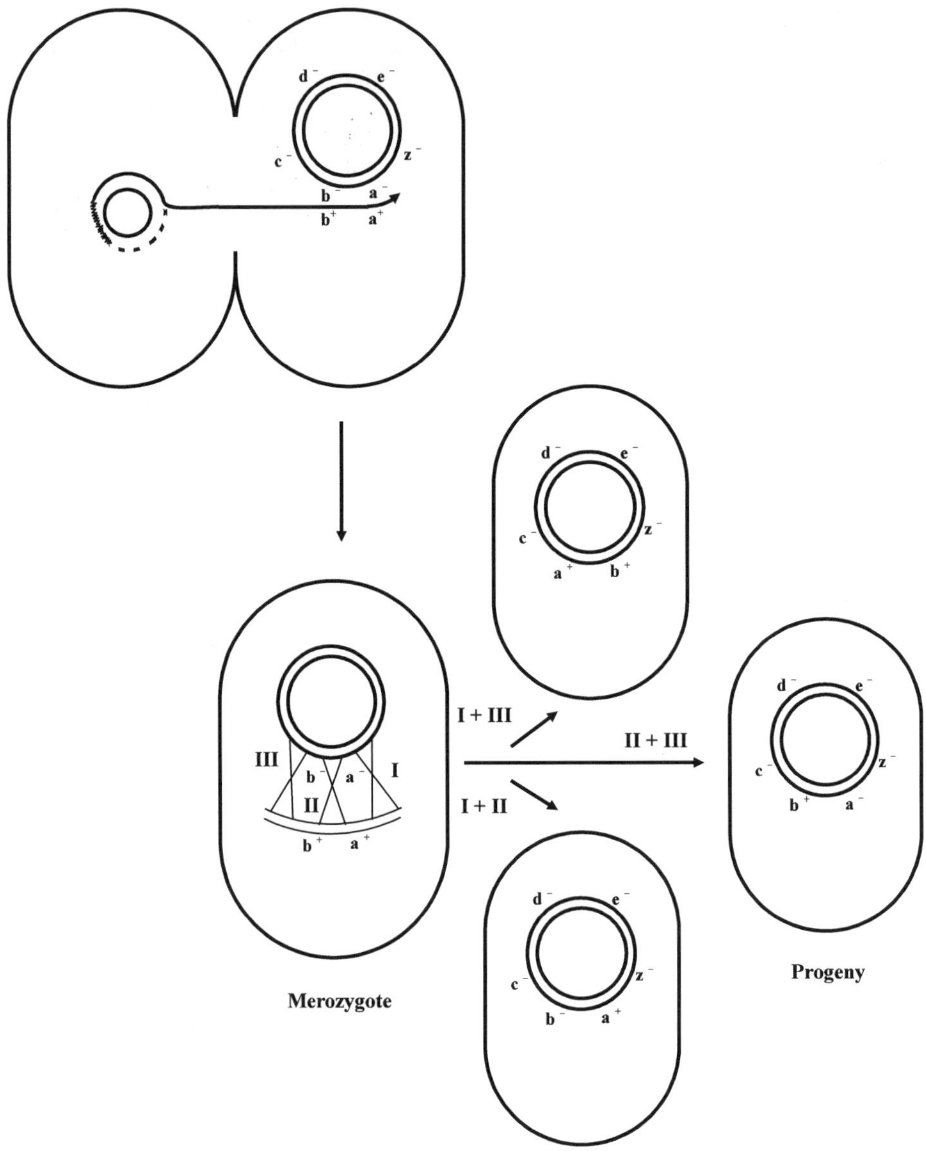

Fig. 3.4 Merozygote formation due to partial transfer of donor chromosome. Recombination with the chromosomal markers of the recipient can lead to different progeny types. I, II, and III represent the possible crossing over events

The prototrophic (a^+ b^+) recombinant will have lac^- as the recipient marker, unless lac^+ is closely linked to a^+ donor marker and thus the two get transferred together. In a reciprocal cross, F^+ lac^- a^- b^+ X F^- a^+ b^- lac^+, however, the a^+ b^+ recombinants will be lac^+ unless lac^- is linked to b^+. Such a problem does not arise if both the parents contribute their complete genetic material as in eukaryotes. The unidirectional and partial transfer of donor genetic material, therefore, must be taken into consideration while using such crosses for linkage analysis.

3.4 High Frequency Recombination Donors

In 1952, Lederberg, Cavalli, and Lederberg and in 1953, Hayes discovered independently a new class of donor strain that had the ability to

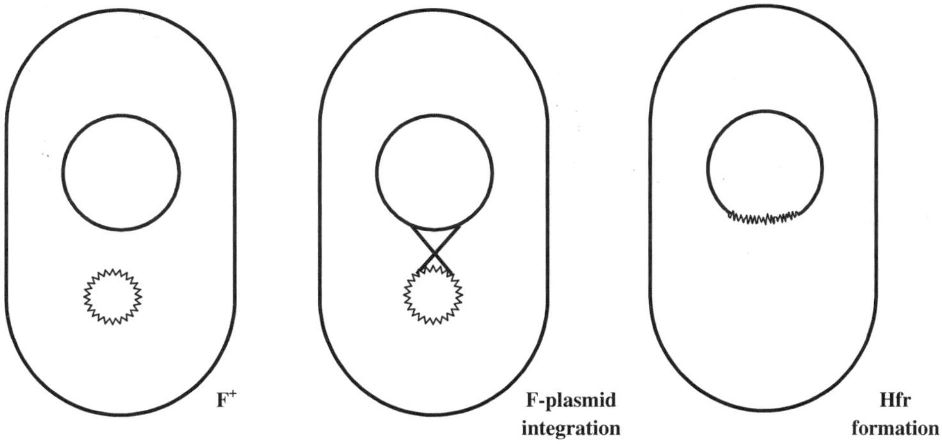

Fig. 3.5 Formation of an Hfr from F^+ strain. The latter carries the free F-plasmid, which can get inserted in the bacterial chromosome to form Hfr. This whole process is reversible thus regenerating F^+ state

promote *h*igh *f*requency *r*ecombination and thus referred to as Hfr. This strain called Hfr H was derived from F^+ 58–161 (*met*$^-$ *bio*$^-$) and when crossed with the standard W677 (*thr*$^-$, *leu*$^-$, *thi*$^-$) produced recombinants at a frequency which was $\sim 10^3$-fold higher than the earlier $F^+ \times F^-$ cross.

Further characterization of Hfr suggested that in compatibility reaction it behaves like F^+ in being able to function only as donor. However, its transfer is not infectious, i.e., the recombinant progeny remain like F^- and thus no conversion generally takes place unlike that in $F^+ \times F^-$ cross. The Hfr nature also appeared to be limited. For example, with Hfr H the transfer of *thr*$^+$ *leu*$^+$*lac*$^+$ show remarkable improvement but that of *met* appeared to fall in the same range as in usual $F^+ \times F^-$ cross. The Hfr can also lose its ability of efficient transfer and revert to an F^+ state, suggesting that Hfr is derived from F^+ (Fig. 3.5).

In a usual Hfr $\times$ F^- cross, generally the transfer of prototrophic alleles from donor to a multiple auxotrophic recipient is followed. One must remember that the principle of counter selection will apply so as to check the growth of the parental strains, as described above and in Chap. 1. This generally involves an antibiotic resistance marker, with the prototrophic donor carrying sensitive and the recipient the resistance allele. The selection medium is designed to select for a donor marker, along with the supplementation of the antibiotic used. Wollman et al. (1956) had suggested that from such an analysis one can determine the frequency of recombination for any selected marker and the genetic constitution of these recombinants. While the former is calculated by estimating the number of recombinants on an appropriate selection agar, the latter is determined by looking at the distribution of unselected marker(s) amongst the recombinants.

3.5 Kinetics of Gene Transfer and Mapping

In order to follow the kinetics of gene transfer, Wollman and Jacob (1955, 1957) devised the technique of interrupted mating as discussed in Chap. 1. Essentially, it consists of mixing the two parental populations and an aliquot of mating mix is agitated on a Waring blender at various times. The mixture is then diluted and plated on selection agar. The agitation does not damage the cells but separates the mating pairs thus stopping the conjugation (transfer of the male genetic material). In some of the earlier protocols, the same effect could also be brought about by diluting the culture at different times after mixing the two strains. Distinct differences can be seen if mating populations are diluted and plated directly with that of when mating is interrupted by agitation. In an

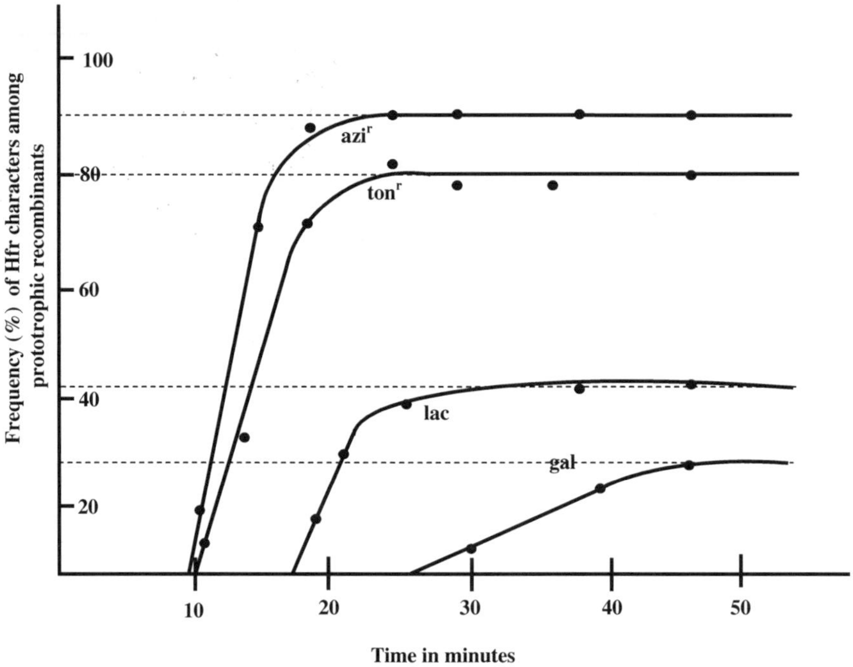

Fig. 3.6 The kinetics of transfer of unselected donor markers in an interrupted mating. Note that each marker appears at a characteristic time and its frequency stabilizes by 50 min

uninterrupted mating, all recombinant classes rise from the origin to reach a plateau at ~50 min. In interrupted mating, however, different types of recombinants appear at different times, though they also reach the plateau level at ~50 min. This difference can be explained by the fact that in non-agitated samples, mating continues until the cells are plated on selection medium and dilution only prevents the fresh pair formation but, on the other hand, the agitation not only separates the mating cells but also prevents any fresh pair from being formed. The time at which agitation is introduced, in fact, will determine the amount of donor material being transferred (Fig. 3.6).

Employing such a technique, Jacob and Wollman were able to demonstrate that each marker from a specific Hfr donor has a characteristic time at which it appears amongst the recombinants. They suggested that this time in fact reflects the time taken by a particular donor marker to enter the recipient cells. This formed the basis of a temporal map of *E. coli* in which genes were located by a unit of time (in minutes) of entry (Fig. 3.7). In an interrupted mating, the frequency distribution of unselected marker will depend upon the time allowed before the mating is brought to a halt. For such an analysis, the recombinants are selected for selected markers and subsequently classified for the presence of unselected donor markers. The frequency distribution of each unselected marker is expressed relative to the number of cells in which conjugation has proceeded as far as the selected markers. An unselected marker also appears at a characteristic time, reaches its maximum value, and then levels off within ~50 min. Thus, if one plots percent of representation of unselected marker in prototrophic recombinants against time there will be a characteristic time at which the transfer of each marker will begin. This identifies the location of that gene on the temporal map. The earlier the gene transferred, the higher will be the frequency it can attain in the population (see Fig. 3.6).

From such an analysis, Jacob and Wollman were able to propose that the Hfr chromosome is transferred to F^- beginning at a defined point, the origin of transfer, also referred to as *oriT*,

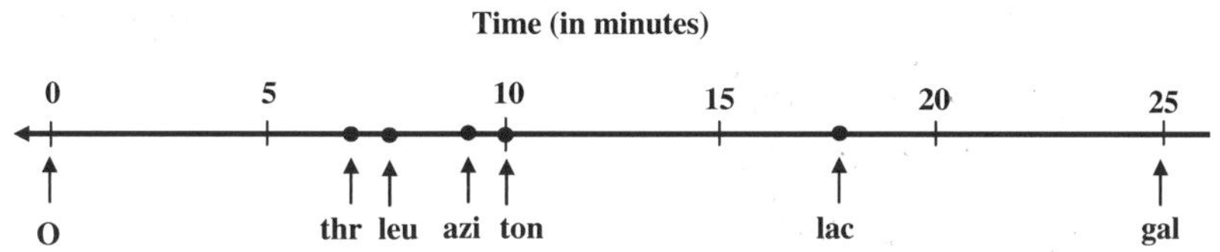

Fig. 3.7 Information obtained from interrupted mating (Fig. 3.6) formed the basis of temporal gene mapping in *E. coli.* The genes are mapped on the basis of time (in minutes) at which they appear amongst the recombinants and temporal units that form the basis for the construction of a linkage map. *Note* The genes are not equidistantly placed suggesting variable linkage between them

followed by the sequence of the other markers in a linear order. The closer a gene is to the *oriT*, the less the time taken for its entry, but those located farther, chances of their inclusion amongst the recombinants (transfer) are remote. In other words, their entry will gradually decline as the distance from *oriT* increases. Genes close to the origin of transfer could be referred to as proximal (early) genes and by the same analogy those located distantly would be called distal (late) genes (Fig. 3.8). The kinetic analysis suggested that bacterial matings suffer spontaneous breakdown of donor–recipient pairs, an effect that is simulated by agitation in interrupted matings. Thus the location of a gene with respect to the origin of transfer is one of the most crucial factors in transfer of that gene, more so because the donor chromosome can only be partially transferred to the recipient. In other words, the probability of transfer of a marker decreases exponentially with its distance from origin of transfer, *oriT*.

Extensive mapping analysis of markers based on time of entry suggested that transfer is linear, a fact that was also corroborated by the effect of temperature on transfer. For example, Wollman et al. (1956) showed that at 37 °C (optimum temperature for growth of *E. coli*), thr^+ leu^+ take 8 min, and lac^+ 17 min to enter the recipient from HfrH after the mating has begun, but at 32 °C, the same two markers took 18 and 32 min, respectively. They opined that though the transfer remains linear it slows down at lower temperature reflecting that the transfer process requires energy. Wollman and Jacob (1958) also showed that many distal markers took a very long time for transfer, which coupled with spontaneous breakdown of the mating, reduced the probability of their presence among the recombinants. The partial transfer of donor chromosome explained the lack of conversion in Hfr × F^- crosses. In fact, the transfer of whole chromosome requires about 2 h making it a very rare event. Subsequent experiments reduced this time to 100 min but transfer of donor type fertility remained a rare event. The modern day temporal map of *E. coli* has, in fact, been standardized on a 100 min scale.

Hfr donors isolated from different *E. coli* K-12 strains by Jacob and Wollman (1957) differed in the sequential order in which the markers were transmitted from donor to recipient, though it remained linear and partial. The comparison of these sequences suggested that all the markers are linked in a single group. The following were the distinctive features of these Hfrs:

1. The *oriT* may be localized at different places in different Hfr strains.
2. Gene/chromosome transfer may take place in either direction (i.e., the polarity of *oriT* may be completely reversed in some strains).
3. Except for the markers close to *oriT*, all markers show the same linkage relationship. In other words, the linkage relationship of two markers will only change if an *oriT* gets positioned between them.
4. Markers located at opposite extremities in one Hfr strain may show close linkage in the other strain.

All these features are depicted in Fig. 3.9a, b. To begin with, these data could not be accommodated in a conventional linear map but a circular map clearly explained all the features of Hfr and its transfer. It essentially suggests that bacterial chromosome is circular but is transferred in a linear manner. The position and

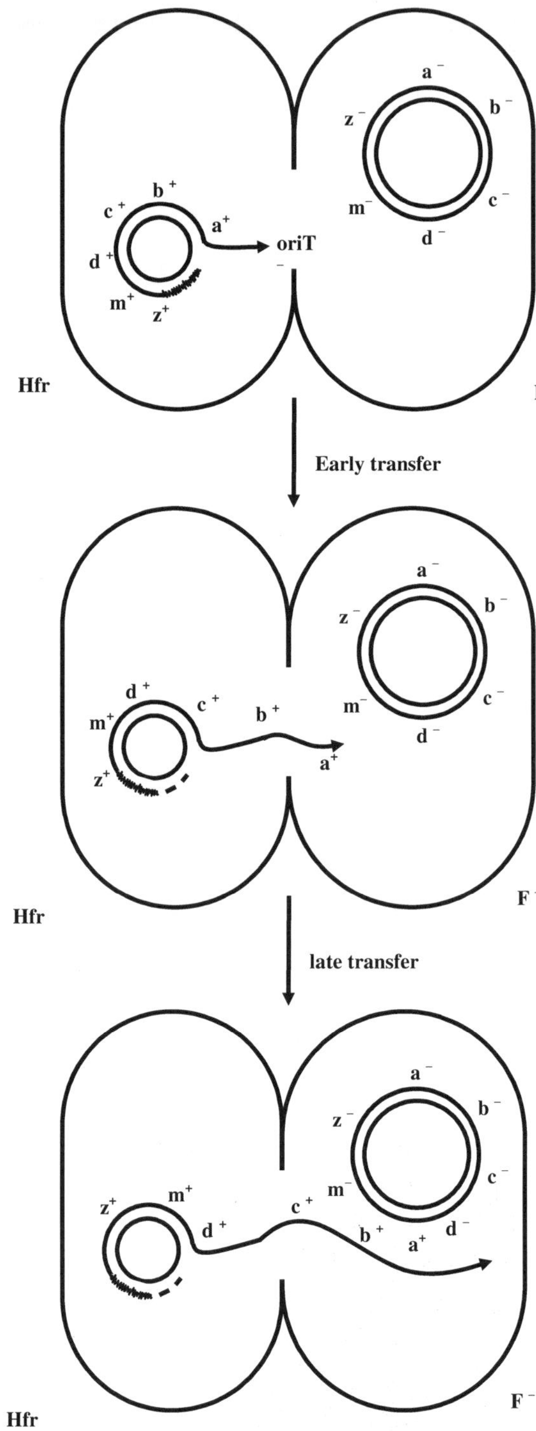

Fig. 3.8 Hfr-mediated time-dependent transfer of donor chromosome markers into F^- cell. Generally, whole Hfr chromosome is not transferred as the mating breaks down spontaneously

orientation of *oriT* in different Hfrs may differ but except for the genes separated by *oriT,* the linkage relationship of other markers does not change. The circular nature of bacterial chromosome was further proved when Cairns (1963) through his autoradiograms of replicating *E. coli* showed it to be a covalently closed circle.

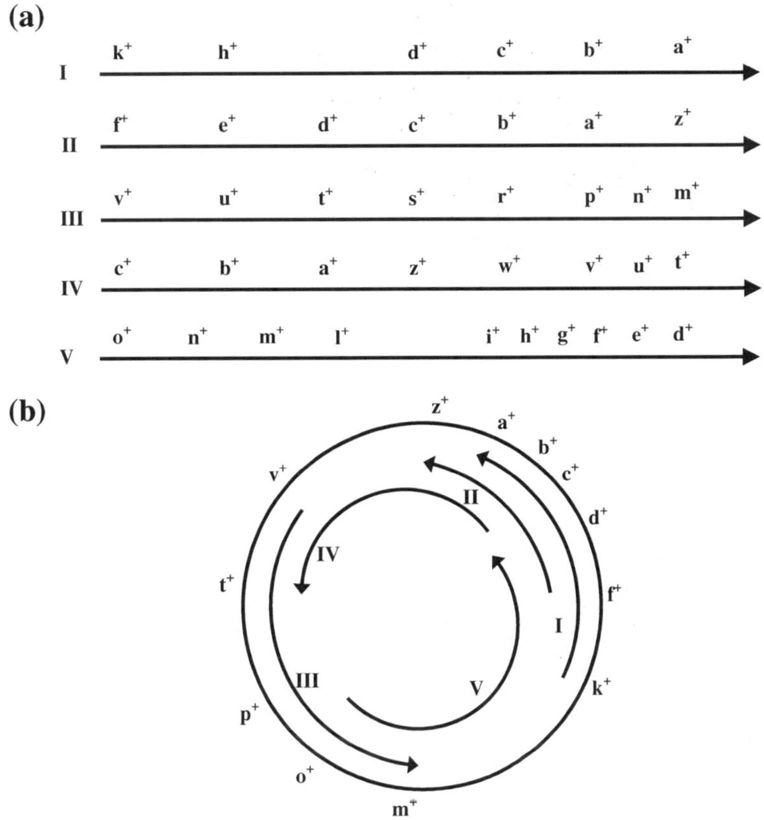

Fig. 3.9 (**a**) Oriented transfer of some hypothetical markers by different Hfr strains. → represents *oriT* in each case and markers are arranged as early and late markers (all markers are not shown). (**b**) This type of transfer can only be explained if the bacterial chromosome is *circular*. *Inner lines* (I–V) indicate the orientation and extent of transfer by different Hfr strains

With the availability of different Hfrs it became possible to undertake extensive mapping exercises in *E. coli.* While temporal mapping remains the most popular method even today, conventional recombination mapping can also be carried out. Two ways by which this analysis was carried out consisted of: (1) selecting for a proximal marker and then classifying them on the basis of distribution of distal unselected markers that might also have been transferred and (2) selecting for a distal marker and then determining the presence of proximal unselected markers that automatically must have entered the recipient. In this case, one can even look for the presence of more proximal recipient marker that will arise as a result of recombination between the markers carried by the donor and the recipient chromosomes. Recombination frequency (RF) can be calculated by counting the number of different types of progeny as shown below, and can be further converted to percent recombination by multiplying RF by 100. The percentage of recombination gives the distance between two loci in map units (1 % recombination = 1 map unit):

Note Hypothetical genes, given in a hypothetical order of entry. Str = streptomycin. No b^- progeny is possible as it is selected against. Other markers can also be analyzed the same way.

$a^+ b^+ c^+ d^+ \, str^s$ x $a^- b^- c^- d^- \, str^r$

↓ select b^+ on medium containing A, C, and D + Str

↓ screen b^+ recombinants for a^+ and a^- markers

$$RF = \frac{(a^- b^+)}{(a^- b^+) + (a^+ b^+)}$$

The sensitivity of this technique is limited to the loci, which are located within 10 min from each other. Thus, the greater the distance, the more they appear to be inherited independently with linkage of about 50 %. With *E. coli* chromosome measuring to $\sim 4.6 \times 10^6$ bp (= $\sim$ 2.8×10^9 Da) and entire chromosome requiring 100 min for the transfer this would put the rate of transfer at $\sim$50,000 bp/min.

distinguished. One must also note that there will be a requirement for additional crossovers because of the partial transfer of donor chromosome that will require its subsequent integration in the recipient chromosome to give rise to the recombinants.

It soon became clear that mere transfer of donor genetic material does not ensure its presence among recombinants. This inference was

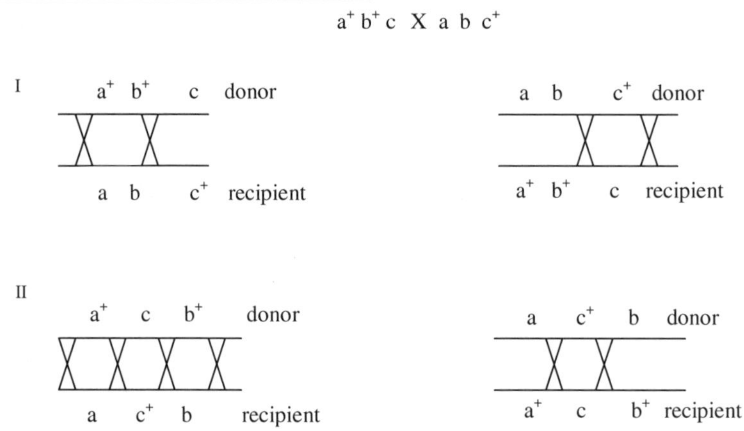

Like eukaryotes, the 3-point crosses can be carried out to determine the relative gene order. This is based on the premise that sometimes the two loci may be so close that it may become difficult to order them in relation to the third, just on the basis of recombination frequency (see Chap. 1). So if the recombination frequency (RF) between a and b cannot be experimentally distinguished from that of a–c, for example, it will be difficult to say whether the order is *abc* or *acb*. In such cases, a pair of reciprocal crosses using the same marker genotypes both as donor and recipient are conducted and prototrophic recombinants ($a^+b^+c^+$) are scored, as shown above.

If we compare the results of two sets of reciprocal crosses, in order *abc,* the prototrophic recombinants will arise at a similar frequency, arising in both the cases by a double cross over. But if the order is *acb,* the first cross will require quadruple and second double crossovers to generate such a recombinant class. As discussed in Chap. 1, the former event will be rare based on frequency (four cross overs) as compared to the other (two crossovers) and can be easily

drawn through a phenomenon referred to as zygotic induction, though not directly related to recombination. Some of the early Hfr strains employed were lysogenic to phage λ (i.e., they carried a copy of phage λ genome integrated in the Hfr chromosome), and which could be transferred to an appropriate F^- through conjugation. However, the progeny type differed in relation to whether Hfr or F^- or both carried the phage λ.

	Type of the cross	Progeny obtained
1.	Hfr × F^-	+
2	Hfr (λ^+) × F^-	–
3	Hfr (λ^+) × F^- (λ^+)	+
4.	Hfr × F^- (λ^+)	+

Note + indicates progeny scored, and – indicates no progeny scored

Of the four crosses shown above, the (1) is the control cross, where both the strains are free of λ and thus will behave normally. In cross (2), on the other hand, as soon as λ is transferred to F^- cytoplasm, it gets induced and 'zygotes' will lyse leading to it being referred to as zygotic

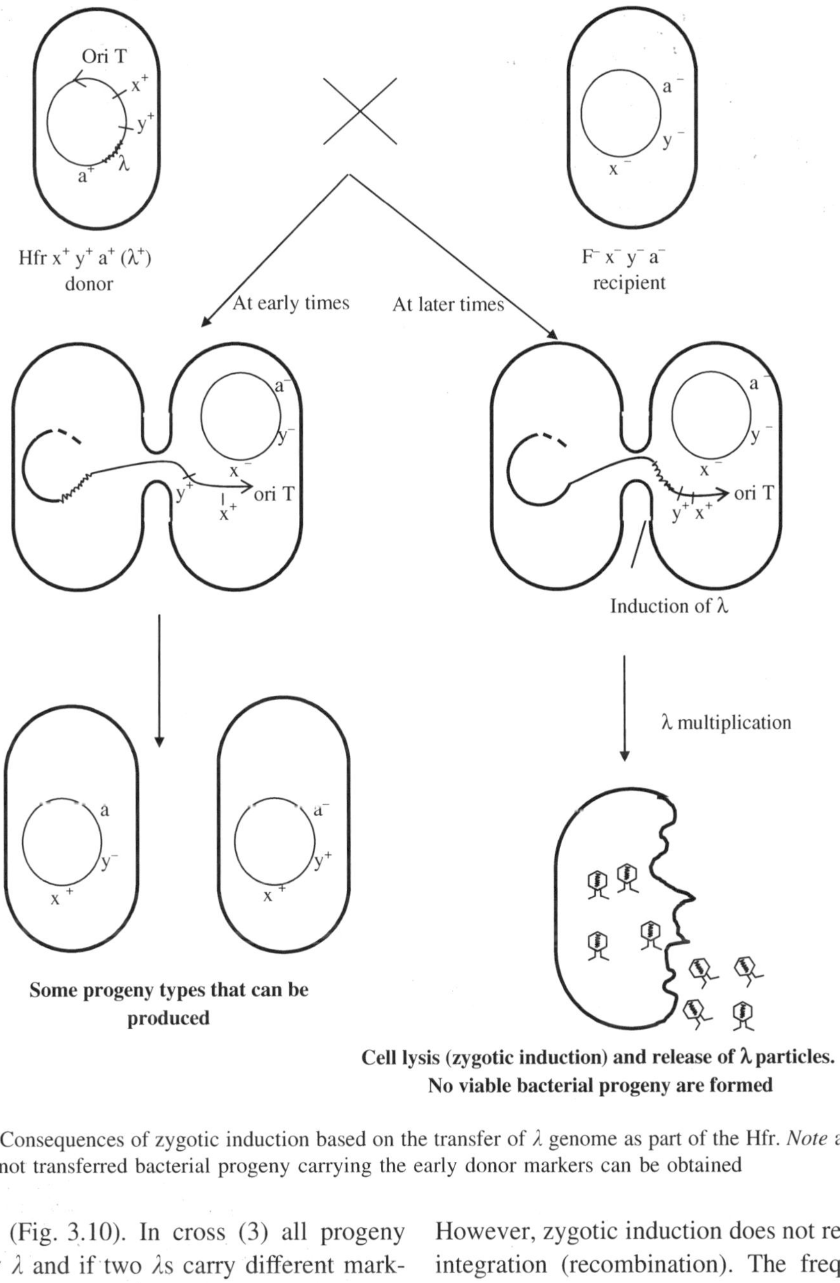

Fig. 3.10 Consequences of zygotic induction based on the transfer of λ genome as part of the Hfr. *Note* as long as λ genome is not transferred bacterial progeny carrying the early donor markers can be obtained

induction (Fig. 3.10). In cross (3) all progeny will carry λ and if two λs carry different markers, recombinants between them can also be formed but progeny will survive. In cross (4), Hfr is λ^- and F$^-$ is λ^+, and thus, λ will segregate into λ^+ and λ^- amongst the progeny.

The occurrence of zygotic induction reiterated the unidirectional transfer of genetic material as it is displayed only when Hfr parent is carrying λ. Moreover, the transfer is linear, as all the markers before λ are transferred normally. However, zygotic induction does not require any integration (recombination). The frequency of zygotic induction can be used to determine as to how many cells are involved in zygote formation and with what efficiency of integration.

Wollman et al. (1956) compared the frequency of *thr*$^+$ *leu*$^+$ recombinants (the early donor markers) in two crosses: Hfr *thr*$^+$ *leu*$^+$ λ^+ X F$^-$ *thr*$^-$ *leu*$^-$ λ^+ and Hfr *thr*$^+$ *leu*$^+$ λ^+ X F$^-$ *thr*$^-$ *leu*$^-$ λ^-. In the first case, prototrophic recombinants are formed at a frequency of 20 %

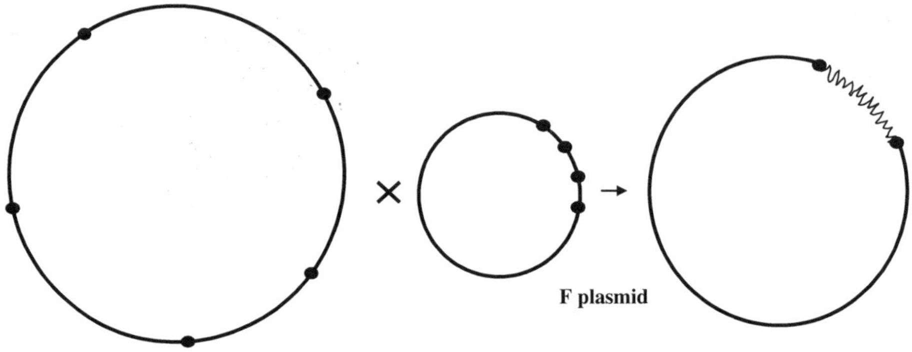

Fig. 3.11 *E. coli* chromosome and F-plasmid carry insertion sequences (●) which are the sites for genetic exchange through which F plasmid gets inserted into the chromosome to generate an Hfr strain. Integrated F plasmid will be flanked by two copies of IS elements

of the parental Hfr number. This in a way sets a minimum value for the frequency of zygote formation. When the results of second cross were scored, 50 % of cells (possible zygotes) got destroyed and the frequency of prototrophic recombinant went down to 10 %.

These results can be explained by the fact that if only 50 % of zygotes are destroyed by zygotic induction another 50 % cells must have survived due to spontaneous breakage of the transfer of donor chromosome. In other words, 100 % of Hfr cells are capable of sponsoring mating. But only 20 % cells are expressing thr^+ leu^+ proximal markers suggesting that efficiency of integration of these markers is only 20 %, or in other words in the majority of zygotes, transfer may not be followed by integration or genetic recombination.

3.6 Generation of Different Hfr Strains

Distinct characteristic features of Hfr strains permitted the speculation on possible pathways of creation of an Hfr. First, the non-transferrable nature of Hfr genotype in contrast to very effective transfer of F^+ suggested that there is some change in the state of fertility factor so as to prevent its transfer in the former state. Moreover, in F^+ this transfer is due to its free autonomous state and obviously in Hfr this may not be free but integrated into the bacterial chromosome (see Fig. 3.5). A specific orientation of transfer also suggested that origin of transfer point is located on F factor and that F factor is integrated in bacterial chromosome at different sites and orientation creating differences between different Hfrs. Reversibility of the formation of Hfr from F^+ with each event occurring at a specific frequency also suggested that it is a single-step process.

From all these features, F factor appeared to behave as a plasmid, which has the capability to exist either in free autonomous state or integrated into the bacterial chromosome (also referred to as episome). Using analogy with Campbell's phage integration model, it was suggested that F plasmid is also circular and can integrate in the chromosome with the help of a single cross over. There are obviously different sites, identified as IS, where integration could take place leading to different Hfrs (Fig. 3.11). The Hfr chromosome must break at the point of *oriT* during transfer. With *oriT* aligned to get transferred first to the recipient the rest of the bacterial chromosome gets dragged along in a time-dependent manner. The major part of the F-plasmid remains at the other end and is rarely transferred to the recipient. First because it requires ~100 min for complete transfer whereas mating may break up spontaneously, much earlier than this, and secondly its non-transfer ensures lack of conversion of F^- or the progeny derived there from to Hfr, a feature normally associated with an F^+ donor (Fig. 3.8).

3.7 F-Prime Formation

While various details of bacterial conjugation were getting revealed, Adelberg and Burns isolated a third type of donor strain. This donor was isolated from a characterized Hfr strain P4x, transferring *pro*, *leu*, *thr* as early and *lac* as a distal marker. They showed that in a cross using subclone of P4x, P4x-1 X F^-, the recombinants gaining early markers also acquired the capability to behave as donors. Chromosome transfer by these donor (recombinants) types resembled that of ancestral Hfr strain, (though at a lower frequency), yet the transfer of sex factor remained independent and infectious like F^+.

This strain thus appeared to have characteristics intermediate between Hfr and F^+ and was referred to as F′ (F-prime). They proposed a model to account for these features and stated that F′ is generated by an imprecise excision of F factor from its integrated Hfr state. Imprecision is brought about by an exchange between a site on the plasmid and another on the chromosome. This type of exchange known as type I exchange generates an F′, in which F plasmid gains some chromosomal genes at the cost of losing some of its own material. It should be clear that the genes flanking the integrated F-factor only, either proximal or distal, can be included in such an imprecise exchange (Fig. 3.12). F′ factor carrying proximal gene/s is referred to as type IA, whereas the one carrying the distal gene\s is known as type IB. Also, bacterial chromosome will carry a part of F-plasmid often referred to as sex affinity site or *sfa*. Since F′ factor is free it can be infectiously transferred. When such a factor is, however, transferred to F^-, it preferentially pairs with the region homologous between F^- chromosome and chromosomal segment carried by F′. A cross over can lead to integration and an Hfr can be generated. The chromosomal transfer by such an Hfr, however, will be at a frequency lower than the original Hfr but higher than that with a conventional F^+. This explains the dual nature of these F′ strains. F′s are labeled along with the gene(s) that they may carry. For example, one carrying *lac*$^+$ will be referred as F′ *lac*$^+$.

When such an F′ strain is generated for a distal marker of Hfr (*lac*$^+$ in the above example), a marker which is otherwise rarely transferred will now be transferred infectiously. Also, when an F′ is transferred to F^- carrying a copy of the same gene, the resulting cell will become a partial diploid and if it gets integrated it will transfer one copy as proximal and another one as distal marker (Fig. 3.13a). Type I exchange generated F′s also carry the only copy of the gene and thus such cells cannot be cured of the plasmid under the selection for the same marker. For example, F′ lac$^+$ cannot be cured of this plasmid on a medium, say in this case, containing lactose.

In subsequent years, another type of illegitimate exchange called type II exchange was observed. In such a case, F plasmid comes out fully but picks up bacterial chromosome portions from both its flanks. Such F′s carry genes from both sides of their site of integration but do not lose any of their own sequences. One may assume that two sites involved in exchange both lie on the chromosome but on the two sides of integrated F (Fig. 3.12). Because a type II F′ factor carries both proximal and distal markers, upon entering a new recipient possessing the corresponding regions it will integrate through double crossover reverting to the Hfr state of the ancestral strain (Fig. 3.13b).

Many F′ factors have now been isolated and many such are sought after by bacterial geneticists because they serve as a very important genetic tool. First, F′ generation confirmed that Hfrs are generated by the integration of F plasmid in bacterial chromosome. Second, such F′s when transferred to F^- cells can generate partial diploids that have been used in dominance analysis of an allele in an otherwise haploid bacterium (see Fig. 3.13a). For example, if an F^- lac$^+$/lac$^-$ partial diploid behaves as *lac*$^+$, this locus is dominant over *lac*$^-$. Also *cis* functions can be distinguished from *trans*- required ones. For example, F′*lacI*$^+$ can normally regulate a chromosomally located (in *trans*) *lac* operon having defective regulator (*lacI*$^-$). But if there is a defective *lacO* or *lacP* mutation that will not be complemented by F′ carrying this region (in *trans*)

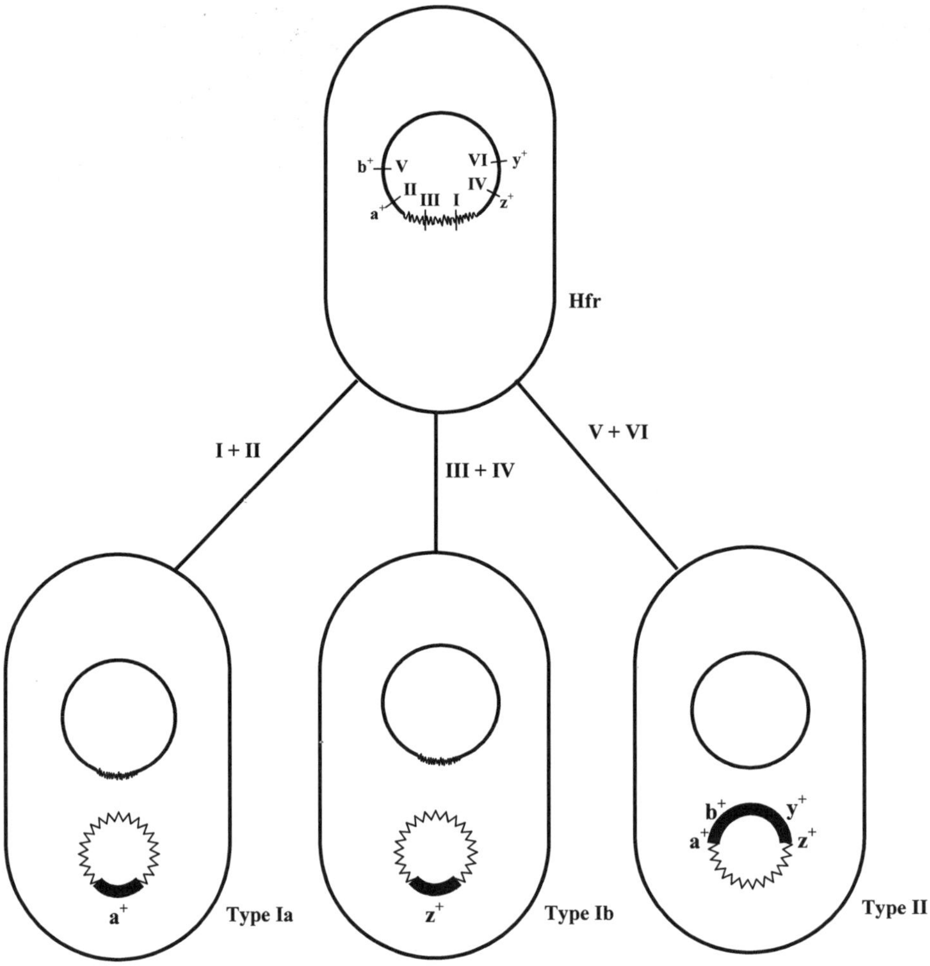

Fig. 3.12 Imprecise excision of F plasmid from an Hfr to generate F′. Depending upon the site of exchange, such as I and II will generate type Ia, III and IV type Ib, and V and VI will give rise to type II F′

showing that these functions are required in *cis*. With such applications of F′ strains they have been regularly isolated. But as they can easily integrate into the chromosome due to homology, they can be best maintained in *recA* cells.

3.8 Structure of F Plasmid

As discussed earlier, F plasmid was assumed to be a circular structure like that of bacterial chromosome. With the advent of improved biochemical and biophysical techniques it became possible to isolate F plasmid free of chromosome. Freifelder (1968) used two techniques to distinguish circular molecule from that of linear ones. These were based on sedimentation properties under appropriate conditions and decrease of density by the intercalation of ethidium bromide (EtBr). Sedimentation behavior of different forms of DNA at denaturing alkaline pH (pH > 11.3) is different. Covalently closed circular DNA is unable to unwind and therefore, sediments three times faster than the linear DNA of the same molecular weight. If a nick is introduced, that will result in one linear and one circular molecule, circle will sediment 14 % more rapidly than linear molecule.

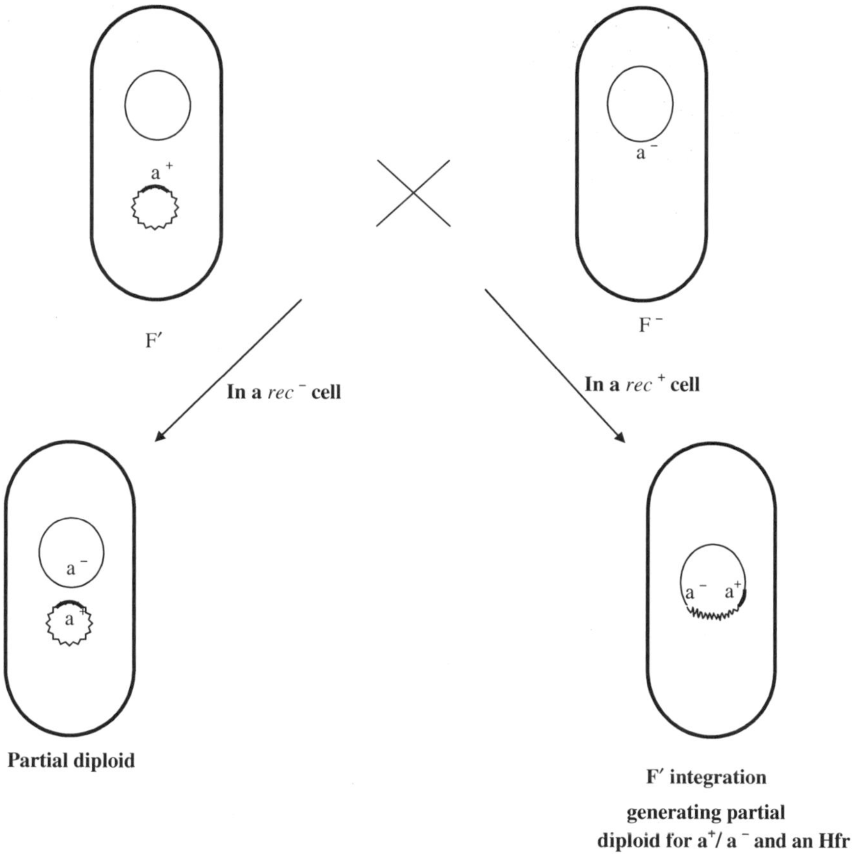

Fig. 3.13a The possible events that may take place when Type I F′ is mated with and F^- cell. When latter is rec^-, a partial diploid will be generated that can be used for genetic analysis. In a rec^+ cell, however, the plasmid with the bacterial gene will integrate into the chromosome, of which one copy will be transferred early and the other late during further mating

The second technique is based on stacking properties of ethidium bromide between the DNA bases. By doing so it decreases the density of DNA by ~ 0.15 g/cm^3. This intercalation involves unwinding, which occurs by reverse twisting of the strands; twisting increases as more EtBr binds. Since in a covalently closed circular molecule two strands cannot fully unwind the strands undergo limited level of twisting, correspondingly less stacking of EtBr, and lesser loss in density. In comparison, in a linear molecule the unwinding takes place easily because there is no constraint on reverse twisting. More EtBr gets bound and therefore, density loss is also more. Thus, on a CsCl density gradient at saturating concentration of ethidium bromide, covalently closed circular DNA will be seen at a denser region in comparison to linear DNA. By using such techniques, several F and F′ were shown to exist as circles.

In order to study its structure and determine some physical attributes, heteroduplex mapping and visualization under electron microscope was resorted to. This mapping is based on the principle that when two different DNA preparations are denatured and then renatured, two types of molecules can be generated. Those formed due to pairing between the strands of the original molecule will produce a homoduplex, but those in which two strands are derived from two different sources will create a heteroduplex. Based on differences in the base sequence of DNAs derived from two sources, a heteroduplex is expected to carry both duplex regions reflecting

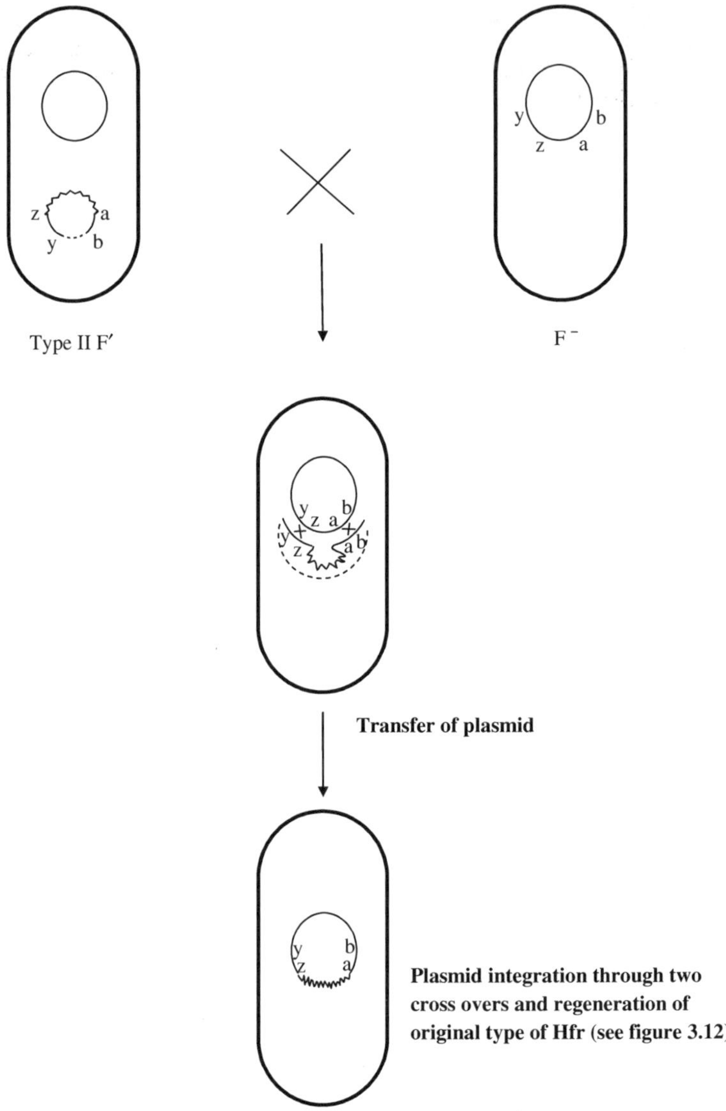

Fig. 3.13b Mating of a Type II F′ with F⁻ will lead to plasmid transfer and its integration into bacterial chromosome through two crossovers that will also generate the original Hfr type. - - - is a hypothetical line to clarify the location of these genes on the F′ plasmid (〰)

the similarities and unpaired single-stranded regions amounting to the differences between them. Homodulex, on the other hand is expected to be uniformly double-stranded structure.

For electron microscopic viewing, DNA preparations are mixed with a basic protein such as cytochrome that gives it a firm structure. Then the preparations can be spread in aqueous medium or in medium containing appropriate concentration of formamide and salt. In the former, single-stranded regions collapse to form bushy structures, but in the latter both appear linear except that the single-stranded regions are thinner in diameter than duplex DNA.

Heteroduplex mapping was employed to look at the structure of F plasmid by comparing a heteroduplex formed between F/F′. F and F′ DNA are nicked, denatured by alkali, and then single strands are reannealed with each other. One information that can be collected is about the type of excision that generated the F′. You may recall that F′ can be formed by type I or type II exchange. In typeII exchange, F plasmid comes out cleanly and picks up extra DNA from

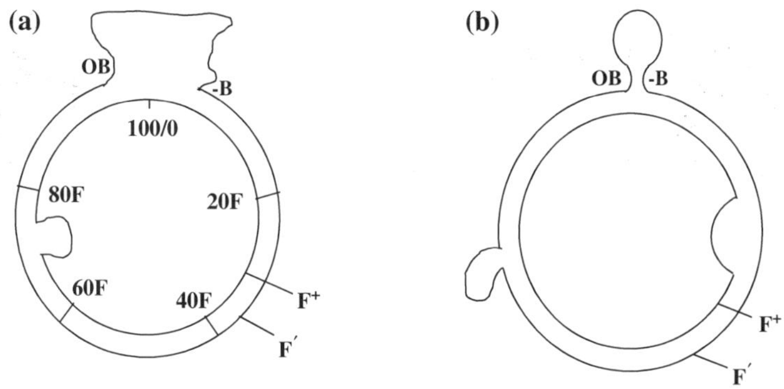

Fig. 3.14 F/F′ Heteroduplex: **a** substitution loop, if the F′ is of type I, and **b** insertion/deletion loop when F′ is type II. Other single-stranded loops can be seen depending upon if F or F′ has suffered other deletions/insertions

the chromosome from both sides of the point of insertion. Such a heteroduplex will show perfect pairing of F sequences between F and F′ but the bacterial DNA will be seen as an insertion loop (also sometimes known as deletion loop). With a type I exchange F′, however, F plasmid gains a part of the chromosome and loses a part of its own. Thus, in a heteroduplex, F′ will show a single-stranded loop corresponding to bacterial sequence and F will also form a loop at the same point as its corresponding sequence is deleted from F′. Such a structure is referred as substitution loop (Fig. 3.14).

Comparison of F/F′ heteroduplexes from different sources as well as those with various reference F′s led to the delineation of a map of F factor. Such a mapping procedure is also referred to as heteroduplex mapping. For such maps, certain co-ordinates are first defined. For example, based on the total contour length of F DNA, the size of the F as well as the length of the inserted DNA could be determined. Also, the length of DNA lost by F during exchange (type I) can be estimated. The junction between the bacterial DNA and sex factor DNA identifies the starting point and F-factor sequences are represented in kilobase pairs (kbp) bearing the suffix F in the clockwise direction. In order to distinguish the chromosomal and the F sequences, the latter bears the suffix F and the former B. Since the sex factor is circular the starting point bears two coordinates representing the starting and end points (O/100 F). The two junction points between bacterial sequences and sex factor are represented as O and B, the latter is prefixed with the length of bacterial DNA inserted.

In order to assign map locations to sequences on the sex factor, F′ factor with distinctive structural features such as small insertions or deletions beside bacterial sequence insertions have also been employed. Deletions and insertions in other factors can be located by heteroduplex mapping. Several deletion stocks of F factor have also been used to create heteroduplexes with reference to F or F′ factors. This helps in demarcating the length of the deleted region and delineating the structure of the F plasmid (Fig. 3.14).

From several such studies some important conclusions could be drawn:

1) Although different F′ tend to carry different lengths of bacterial DNA, many of them appear to use a hot spot where recombinational excision (and therefore integration) takes place (0–2.8F).
2) All integration/excision events seem to be restricted to 0–15.0F. The loss of F DNA in this region during excision did not affect the viability of F′ strains generated, leading to this region being referred to as non-essential or dispensable.
3) Comparison of the sequence and orientation of gene transfer by different Hfrs suggested that F plasmid has defined sites for integration into the bacterial chromosome. All integrations also involve specific regions of the F factor itself.

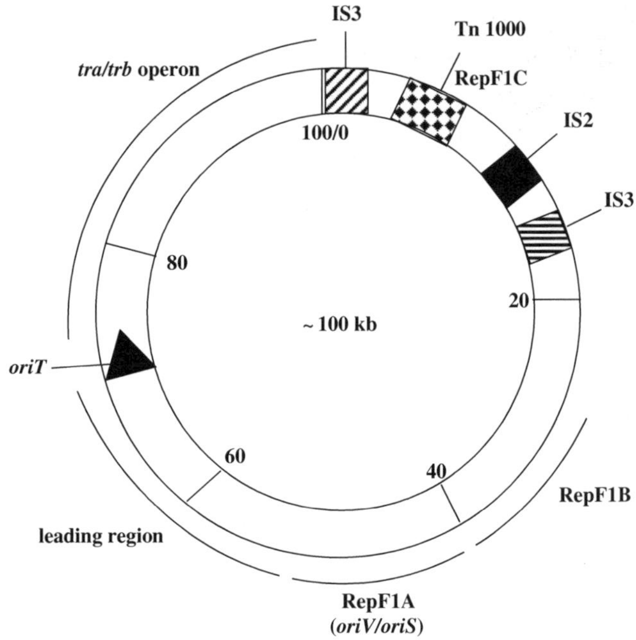

Fig. 3.15 Structure of F plasmid showing important components (For details see Fig. 6.16)

The sites for recombinational insertion of F DNA were identified through several studies. In one such study, information was derived from a class of strong polar mutations first identified in *E. coli galT* gene. Such polar mutations were different from the known polar mutations in having an increased density and the shifting of density to normal range upon reversion. This led to the suggestion that these unusual polar mutations are caused by the insertion of DNA into *galT* gene. Several such insertion mutations were subsequently isolated in different genes and their regulatory regions. Interestingly, in these mutations the length of the inserted DNA fell into defined categories, which were labeled as insertion sequences (IS). These were classified on the basis of their length into IS1, IS2, IS3, etc.

The role of these IS elements in integration/excision pathway was suggested when they were found located on bacterial chromosome, F and R plasmids and several phages, more importantly at sites that were likely to be implicated in recombination events. Not only the location, but also the position of the origin and direction of chromosomal transfer by several Hfrs could be explained on the basis of integration of F factor in a particular orientation through an exchange between IS sites of plasmid and chromosomal DNA.

From the different evidence collected so far, including the direct visualization of F/F′ heteroduplexes, F is a large covalently closed circular molecule, measuring to a length of ~100 kb (Fig. 3.15). The plasmid has three origins of replication Rep F_1A, Rep F_1B, and Rep F_1C of which Rep F_1A is the primary origin for replication and includes both unidirectional (*oriS*) and bi-directional (*oriV*) replication origins. Rep F_1B is used only if Rep F_1A is non-functional and Rep F_1C is a non-functional origin due to insertion of $\gamma\delta$ or Tn 1000. Another site, *oriT* is used as an origin for F DNA transfer, as discussed earlier. The region beginning *oriT* is the first to enter the recipient and has, therefore, been referred to as leading region. The nicking site at *oriT* exhibits strong sequence conservation. This site lies on the complementary strand and is flanked on the left side by 40 nt AC-rich and on the right by ~100 nt AT-rich region. The *oriT* region contains a number of direct and inverted repeat sequences. The

AT-rich segment contains intrinsic bends as well as binding sites for IHF and for TraY and TraM. Also contained in this region is the nick site where TraI binds and generate single-strand nick, assisted by TraY and IHF. In the region 0–20F, F plasmid carries two copies of IS3, one copy of IS2, and one copy of insertion sequence $\gamma\delta$ or Tn1000. The huge region of $\sim$34 kb on F is occupied by *tra/trb* operon, the constituent genes of which are responsible for the various functions required for conjugation. These include the synthesis of pilin and assembly of functional pilus, pairing and stability of mating aggregates, triggering the transfer and mediating the actual transfer, nicking at *oriT* and unwinding of DNA, as well as some regulatory genes. The *tra/trb* operon genes code for the pilin subunit (*traA*); pilin maturation (*traQ*, *-X*); the core T4SS proteins involved in pilus assembly and transport (*traB*, *-C*, *-E*, *-G* [N-terminal region], *-K*, *-L*, *-V*), the essential F-specific genes (*traF*,*-G* [C-terminal region], *-H*, *-N*, *-U*. *-W*, *trbC*); auxiliary F-specific genes (*traP*, *trbI*, *orf169*); the regulatory genes (*traJ*, *-Y*, *finO*, *-P*); DNA processing (*traI*, *-M*, *-Y*) and transport (*traD*); mating pair stabilization (*traG*, *-N*) and exclusion (*traS*, *-T*); as well as many other genes whose function is not clearly understood (*traR*, *trbA*, *-B*, *-D*, *-E*, *-F*, *-G*, *-H*, *-I*, *-J*).

The proteins that form a large membrane-spanning complex, called a type4 secretion system (T4SS) provides the channel that is used for DNA transfer. F plasmid is a conjugative plasmid and besides regulating the formation of pili, and conjugal transfer also controls plasmid copy number, incompatibility, maintenance, fertility inhibition, and surface exclusion as also the inhibition of replication of certain female-specific phages. In most F-like plasmids the *tra* genes are negatively regulated by the FinO protein and *finP* RNA (*fin* standing for fertility inhibition). F factor, however, exhibits unhindered transfer due to insertional inactivation of *finO* gene by an IS3 element. The transfer region encodes two genes responsible for limiting the capacity of a cell containing F plasmid to act as recipient, a phenomenon known as surface exclusion. The *traS* gene appears to block the transmission of a mating signal and its 16.9 kDa product is lodged in the inner membrane. The second gene involved is *traT*, whose 26 kDa lipoprotein product is a major outer membrane component of F-containing cells. This inhibits the formation of mating aggregates. We shall look at the details of the structure of F plasmid and its constituent genes also in Chap. 6 on Plasmids.

3.9 Structure of the DNA Transfer Apparatus

The ability of F^+, Hfr, and F' to act as donors is ascribed to certain specific properties acquired by the cell as a result of the possession of F plasmid whether in a free or integrated state. One early difference described by Loeb (1960) was based on the observation that certain bacteriophages lyse only those *E. coli* that had the ability to act as a donor. Such male selective phages include RNA phages F2, R17, MS2, M12, Qβ, and single-stranded DNA phages such as f1, fd, and M13.

The characteristic surface property of the donor strains has been identified as the possession of F pili or sex fimbriae. While all *E. coli* cells are surrounded by hairy appendages called type I pili, donor cells are endowed with an additional type of pilus, the F pilus (Fig. 3.16). The male-specific phages adsorb either to the length (usually RNA phages) or to the tip of the F pilus (DNA phages). The F pili are 1–2 μm long and grow perpendicular to the cell surface. Their number varies with the growth conditions (usually one to five). F pilus is a thick and flexible, hollow rod-like structure with an external diameter of 85 Å and an axial hole of 20–25 Å in diameter. Each pilus is assembled from many subunits of a single polypeptde, F pilin, each $\sim$11.8 kDa in size, with an isoelectric point of pH 4.5, and arranged in a helical array. The plasmids belonging to the same incompatibility group share the pilus structure and its serological properties. As discussed earlier, the production, organization, assembly, and

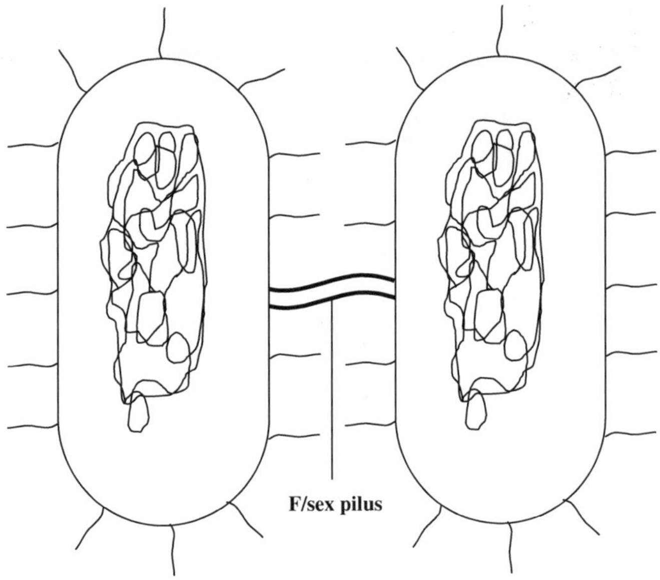

Fig. 3.16 Conjugating *E. coli* cells are seen connected with a long tubular structure called F or sex pilus. Both cell types may be covered with other type of pili, but sex pilus is produced only by the donor cell

function of pilus is governed by a large number of genes located on the plasmid.

The mechanical removal of F pili or addition of anti-F antibodies disrupts the donor activity, which confirms its essential role in conjugation. Electron microscopic pictures also clearly show conjugating *E. coli* cells connected with F pilus. All this leads to the suggestion that F pilus provides the conduit through which the DNA travels from donor to the recipient cell. Whether the pilus provides the channel for transfer of DNA has remained a matter of intense research, some claiming that no direct evidence exists for the presence of DNA into the pilus.

Pili are implicated in as many as three processes: signal transduction, cell–cell aggregation, and DNA transport, though evidence for each one of them may be more indirect. Inability of such donor cells that lack the pili or are prevented from contacting the recipients, to initiate conjugation-specific events led to the suggestion that pili sense the contact between the conjugants and in fact trigger such an event. Shortly after the contact, the pili are depolymerized and the subunits return to the donor cell membrane, a process known as pilus retraction. Pilus depolymerization may help in drawing the donor and the recipient cells close to each other and facilitate the cell aggregate (consisting of up to 80 cells) formation. Mating pairs are stabilized within these aggregates prior to DNA transfer, but intact pili are not required for DNA transfer. Also, if the pili are dissociated by treating with detergents, such as SDS, after mating pair stabilization but before DNA transfer, the transfer process is not inhibited. Thus, pilus base which survives the detergent treatment may provide a channel sufficient for DNA transit. Pili are thus proposed to gather the membrane component of the transfer apparatus and establish contact with the cytoplasmic membrane of the recipient. Others have, however, shown that DNA can be transferred from the donor to the recipient even when the mating cells were apparently physically separated with the only likely contact maintained through the pilus. More recently, fluorescent tagging and the visualization of DNA transfer and recombination by fluorescence microscopy in real time and at the level of individual living cell has provided further support to this view. This study has shown that F pilus, in addition to establishing the contact between mating cells, serves as the conduit for single-stranded DNA transfer.

Many of the *tra/trb* genes contribute to a T4SS, as described above that propels the donor DNA toward the recipient, which in fact is a multistep process.

Although all donor strains possess sex pili, F^+ and F′ strains can be easily distinguished from Hfr on the basis of treatment with agents like acridine orange or acriflavine. Hirota (1960) reported that upon acridine orange treatment, an F^+ strain can be converted into the F^- state, and the recipient status is stably inherited by its progeny. This compound showed no effect on Hfr strains suggesting that such a treatment leads to the loss of free autonomous factor but not of the integrated one. This phenomenon called plasmid curing is based on the inhibition of replication of autonomous sex factor and emergence of plasmid-free segregants in the process. Application of compounds like acridine orange provides a standard test to determine whether the cell carries F or any other plasmid in free, autonomous (susceptible to curing) or integrated (non-curable) state.

While many functions both unique and specific for donor activity were recognized, such functions could not be revealed in recipient strains. It was expected that recipient cells actively participate by providing conjugal receptor on their surface. Although mutants of two classes: (i) those with altered LPS in outer membrane and (ii) those defective in the product of outer membrane protein (*ompA*) were studied, their conjugation defect was more indirect. Such mutants called *con*$^-$ could not identify a single form of LPS or the *ompA* gene product that may be involved in pilus receptor formation. The Con$^-$ phenotype was explained on the basis of an exclusion of the donors from the aggregates. The broad host range for DNA transfer also goes against the existence of a specific pilus receptor.

3.10 Chromosome Transfer and Recombination

As described earlier, the transfer of bacterial chromosome from donor to the recipient is under the control of integrated sex factor and commences at *oriT* located on the F factor. One requirement, as you may recall, is that a nick is introduced in the circular chromosome so as to transfer it linearly.

In 1963, Jacob, Brenner, and Cuzin formulated two classes of models for chromosome transfer. In one, the chromosome is replicated prior to its transfer so that a copy is transferred to the recipient. This model, however, could not explain the nature of the signal required to initiate this replication.

The second model required the concomitant replication of the chromosome while the transfer is on. Although the original model envisaged the transfer of one of the replicas of the donor chromosome, it could subsequently be applied to a situation in which only one of the strands of the donor DNA is freed by replication for its passage to the recipient. Under both the situations, the replication is under the control of sex factor and is presumably activated by cell–cell contact.

Evidence collected from different lines of studies clearly suggest that only one donor DNA strand is transferred to the recipient by all the three donor types: F^+, Hfr, and F′. Such a transfer could be explained on the basis of rolling circle replication (Gilbert and Dressler 1968). This involves the nicking of DNA in one strand at *oriT*, allowing the generation of a free 3′-OH and a 5′-P end. While 3′ end is used for the addition of deoxyribonucleotides to maintain strand growth in 5′–3′ direction, the freed 5′ end is displaced and could enter the recipient. It is reasonable to suppose that the same origin of transfer is employed by all the donor types, free or integrated. In the F plasmid system, endonuclease activity for nicking is associated with the *traI* and assisted by *traY* gene product. These two proteins have also been implicated in orienting the plasmid and the nick site for transfer. TraY, which is located in the cell envelope, alone or in conjunction with TraD, may anchor the plasmid to the cell membrane in alignment with the transit apparatus. TraI has been characterized as a DNA helicase and a DNA-dependent ATPase that remains bound to the 5′ end of the single-stranded donor DNA. The

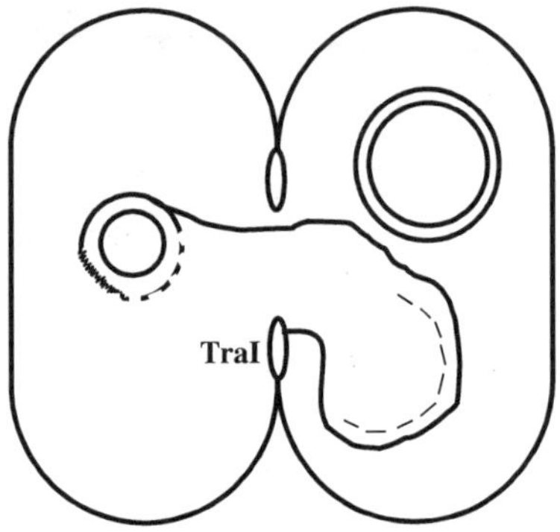

Fig. 3.17 Proposed model for donor DNA transfer to recipient. This DNA is single-stranded and is transferred in the form of a loop where 5′ end remains associated with TraI. The lagging strand synthesis takes place to convert it into a double-stranded structure

ATPase is supposed to provide the energy to propel the 5′ end of the donor DNA into the recipient.

Donor transferred DNA provides a template for lagging strand synthesis in recipient. It has been proposed that the leading 5′ end probably remains attached to the helicase at the site of DNA transfer/replication so that in the recipient it appears as a growing loop of partially duplex DNA (Fig. 3.17). In order to process this donor DNA into viable recombinant products, the function of over 20 genes are summoned by the recipients.

When the mating terminates, the transferred DNA is released as a linear fragment with a ~40 kb region of leading region of F, and a single-stranded overhang of variable length at 3′ distal end. In order to produce viable recombinants, the transferred fragment and the circular recipient chromosome must undergo exchange. The underlying process has been delineated by characterizing a large number of mutants affecting the genes involved.

After the initial sequence-independent interaction, the molecules move in conjunction to align by homologous sequence pairing (synapsis). One can presume that for recombination to occur one of these two DNA molecules must have a single-stranded region that will be utilized by RecA protein to polymerize. This helical RecA-DNA filament then proceeds by strand transfer to be followed by strand exchange. The multifunctional RecA protein thus plays the most crucial role. A minimum of two crossovers (or higher number) will be required to produce a viable circular recombinant chromosome. Smith and co-workers suggested that in 80 % of the matings, these crossovers are initiated by RecBCD enzyme and are located near each end of the transferred Hfr. This multifunctional protein complex, product of the *recB*, *recC*, and *recD* genes, has the ability to load onto the duplex DNA molecules with or nearly flush ends. It has DNA helicase and nuclease activities so that it unwinds and degrades both strands as it travels through the molecule. Its nuclease activity is modulated as it encounters an 8 bp sequence, 5′-GCTGGTGG-3′, referred to as *chi* site so that the strand ending 3′ is no longer degraded. The single-stranded DNA so generated can be recruited by RecA to initiate exchange. Such *chi* sites, considered to be recombinational hotspots, are known to occur every 5 kb on an average on the bacterial chromosome. The next step in this process requires the removal of single-stranded overhang beyond the crossover by a single-strand exonuclease, like exonucleaseI, the *sbcB* product, or RecBCD itself, as well as the leading end so as to expose the homologous sequences between Hfr fragment and recipient chromosome. The latter function is carried out by RecBCD enzyme.

If RecBCD enzyme initiates precisely two exchanges, one near each end, the Hfr DNA will be integrated as a single fragment. Such a process, referred to as long chunk integration, holds good for some 80 % of the recombinants derived. The remainder seems to arise from short chunk integration, where both the crossovers are located near the selected marker. It has also been suggested that recombination could also be initiated by making use of transient single-strand gaps in the loop of Hfr DNA during transfer.

In all cases, RecA-mediated strand exchange results in the formation of Holliday junction. The RuvAB and RecG proteins bind to Holliday junction and catalyze what is known as branch migration. Depending upon the direction, this

Table 3.3 Genes involved in homologous recombination

Gene	Map location (in minutes)	Protein product and the assigned function
recA	60.8	38 kDa: essential for genetic recombination; binds single-stranded DNA; DNA-dependent ATPase; promotes base pairing (assimilation); control of SOS system
recB *recC*	63.6 63.7	135 kDa and 125 kDa: ssDNA and dsDNA exonuclease (part of ExoV); *chi*-specific endonuclease; DNA helicase; ATPase; initiator of recombination by unwinding a dsDNA molecule and creating a region of single-strandedness in an area containing a *chi* site
recD	63.6	55 kDa: exonuclease, part of ExoV
recE	30.5	140 kDa: 5′ to 3′ dsDNA exonuclease, ExoVIII
recF	83.6	40 kDa: binds ss DNA and ATP; can substitute for RecBCD
recG	82.4	76 kDa: binds to Holliday junctions; acts as ATPase; involved in branch migration and resolution of Holliday junction
recJ	65.4	60 kDa: 5′ to 3′ ssDNA exonuclease; can substitute for RecBCD, ssDNA-specific nuclease, also involved in plasmid recombination
recN	59.4	60 kDa: binds ATP; can substitute for RecBCD
recO	58.2	31 kDa: binds ssDNA and promotes renaturation; can substitute for RecBCD; helps RecA to bind to ssDNA; also involved in DNA repair
recQ	86.5	74 kDa: dsDNA helicase can substitute for RecBCD
recR	11.0	22 kDa: binds dsDNA and ATP; can substitute for RecBCD
recT	30.4	33 kDa: ssDNA binding and renaturation, promotes recombination by DNA strand invasion; RecA-independent ds break repair with RecE
ruvA	41.9	37 kDa: binds to Holliday junction; involved in branch migration
ruvB	41.9	22 kDa: binds to RuvA; ATPase; involved in branch migration
ruvC	41.9	19 kDa: endonuclease-specific for Holliday junction; involved in resolution

Note many suppressors have also been identified *sbcB, sbcC* suppress *recBC* mutation and *sbcD* could act as co-suppressor; *rusA* encodes for a *ruv* suppressor

may lead to the extension of the heteroduplex joint or abortion of the exchange. Finally, RuvC, an endonuclease acts as a resolvase to resolve the Holliday junction to patch (non-crossover) or splice (crossover) product. Readers can obtain the details of the recombination mechanism from any book dealing with Genetics, as the available models have been formulated on the basis of information gathered from studies on *E. coli.*

A large number of genes have been identified on the basis of mutants affecting the process of recombination. They are found dispersed all over the *E. coli* genome except for those that form a polycistronic organization such as *recBCD*. In the majority of cases, the function of the gene product has been identified and their role illucidated in the recombinational pathway (Table 3.3).

3.11 Conjugation in Other Gram-Negative Bacteria

The F plasmid can replicate only in enteric bacteria, so the type of conjugation described above can occur only in this group. But a large number of plasmids called conjugative plasmids capable of self-transfer exist in other Gram-negative bacteria. Self-transmissible plasmids have also been described from several strains of *Salmonella typhimurium* but none have been exploited to provide useful genetic information. F plasmid transferred from *E. coli* to *Salmonella* could be used only if resident fin$^+$ plasmids were cured. From such cured strains, Hfr-like variants have been isolated. In some strains of *S. enterica* serovar *Typhimurium* (earlier *S. typhimurium*), F factor gets inserted at one site of the chromosome that leads to chromosomal mobilization. Similar insertions have also been effected by the transposon Tn10 carried both by F and the chromosome.

All conjugative plasmids cannot readily mobilize the transfer of the chromosomal DNA. Some of such plasmids can acquire this property by gaining small pieces of DNA, such as insertion sequences. An example of this is found in plasmid R68 that itself transfers at a high frequency, but mobilization of another plasmid or chromosomal DNA takes place very inefficiently. Derivatives of R68, such as R68.45, have been isolated which mobilize both chromosomal and plasmid DNA in *Pseudomonas aeruginosa* and a variety of other Gram-negative bacteria at a high frequency. The difference between R68 and R68.45 is that the latter contains a 2.10 kb insertion sequence designated IS21.

As described earlier, even the F plasmid owes its chromosomal mobilizing activity to its complement of insertion sequences. However, the mechanism of F-mediated and R68.45-led mobilization appears to differ. The insertion sequences found on F are the same as those located on the chromosome and mobilization requires homologous recombination between these regions, while IS21 sequence may not be found on the chromosome of those bacteria where R68.45 can function. Mobilization in such a case is perhaps a consequence of transposition. Self-transmissible plasmids R18 and R91-5 of *P. aeruginosa* carry Tn1 that can mobilize the chromosome from different Tn1 insertion points. Hfr-like strains have also been described in this bacterium that carry plasmids called FP. They carry out a similar function in this bacterial species as described in *E. coli.*

Agrobacterium tumefaciens is a plant pathogen that induces crown gall tumors in several species. This is brought about by the transfer and integration of an oncogenic piece of DNA, the T-DNA of Ti plasmid into plant chromosomes. Such tumors secrete opines that are not only utilized by the bacterial cells as a nutrient source but also act as a chemical signal to effect conjugal transfer of Ti plasmids to non-Ti plasmid-containing cells. Two types of Ti plasmids are known. While octopine-type of plasmid is induced by octopine, the nopaline-type plasmids respond to agrocinopines (Fig 3.18a, b).

In octopine-type Ti plasmid, the catabolism of octopines and conjugal transfer are coordinately regulated by an activator, OccR. The latter function requires the expression of *traR*, a LuxR homologue of *Vibrio fischeri*, and is a part of 14-gene, octopine-inducible operon, *occ*. The octopine released by the crown gall cells induces the expression of *occR*, which in turn stimulates the expression from *occ* operon including *traR*. TraR alone cannot activate the genes *(tra* and *trb*) required for conjugation, unless it responds to a signal, N-(3-oxooctanoyl)-L-homoserine lactone (OOHL). OOHL is synthesized by *traI*, a close homologue of *luxI* and is located on Ti plasmid. The whole system operates through cell-density-dependent manner based on auto-induction, and thus, is a good example of quorum sensing (Fig. 3.18a). In order to complete the similarities between *lux* operon of *V. fischeri* and *A. tumefaciens* Ti plasmid conjugation, *tra* boxes similar to *lux* box have been found upstream of three operons regulated by TraR, *traAFB*, *traCDG*, and *traI-trb*. Another homologue of LuxR has also been found which responds to mannopines. This has led to the suggestion that *A. tumefaciens* may modulate its conjugal transfer pathway in a nutrient-dependent manner.

The conjugal transfer of nopaline-type plasmid is also regulated by quorum-sensing. In this system, a repressor, AccR represses the action of TraR, which is a component of Ti plasmid-based *arc* operon. This repressive action of AccR is relieved in the presence of agrocinopines, thereby allowing the expression of *traR* and the rest of the *arc* operon. TraR is then able to induce *tra* and *trb* genes in a manner analogous to octopine-type Ti plasmid, that is, only in the presence of OOHL. OOHL is synthesized by *traI* (Fig. 3.18b).

An anti-activator of TraR called TraM has been described both from octopine-type and nopaline-type Ti plasmid that interacts with TraR in an OOHL-independent manner. This prevents TraR to bind to its target promoter sequences. The physiological function of TraM

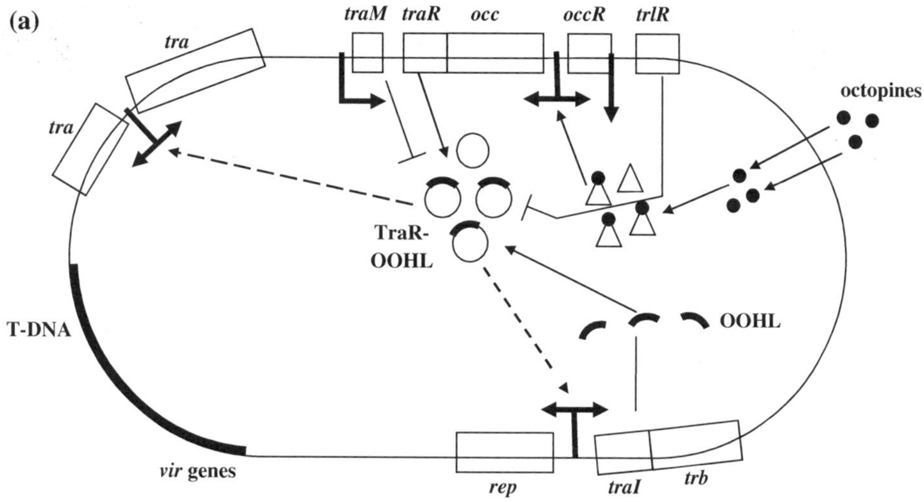

Fig. 3.18a Quorum sensing-based regulation of octopine-type Ti plasmid transfer in *Agrobacterium tumefaciens* (for details refer the text).– – – – – – ➔ represents the *tra–trb* operons activated by TraR-OOHL complex that facilitates the interbacterial transfer of the plasmid. Also shown in the figure is T-DNA with *vir* genes required for plant transformation

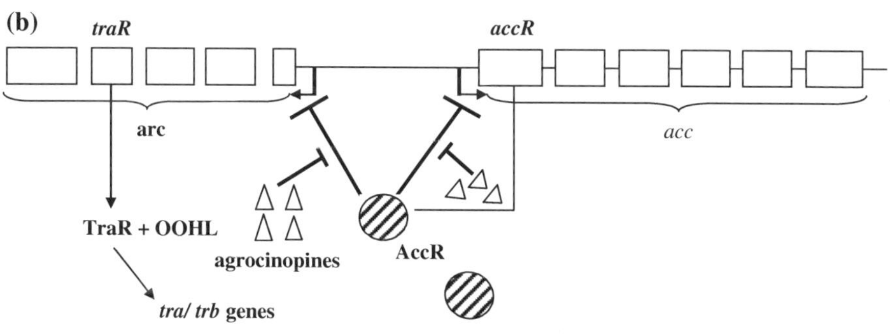

Fig. 3.18b Nopaline type of plasmid is made of two divergently transcribed *arc* and *acc* operons. The transcriptional repressor AccR represses both the operons and the repression is relieved in the presence of agrocinopines A and B. Tra R encoded by *arc* operon is involved in quorum-sensing-based plasmid transfer by interacting with OOHL produced by cognate *traI* gene as shown in opaline type of plasmid

may be to disallow the conjugal transfer at low cell density and tone down quorum sensing after a particular time period. It is believed that the regulated transfer of Ti plasmid may have evolved to ensure that the ability to catabolize a particular opine spreads rapidly in a given population. The reliance on quorum-sensing pathway for conjugal transfer may have been based on the inefficient inheritance of Ti plasmid by the recipient agrobacterial cells. Enhancing the number of Ti donor cells via quorum sensing may brighten up the chances of successful transfer of Ti plasmids.

3.12 Conjugation in Gram-Positive Bacteria

Conjugation as a mode of genetic exchange is less documented in Gram-positive bacteria and where it is known, it is more complicated than Gram-negative bacteria. Self-transmissible plasmids have been reported in many Gram-positive members, such as *Bacillus, Streptococcus*, *Staphylococcus*, and *Streptomyces,* etc. Although the transfer process has not been described in detail they do seem to carry a much smaller *tra* region as compared to Gram-negative bacteria. The plasmid pSN22 of *Streptomyces nigrificans* carry a 7.0 kb *tra* region, coding for approximately five genes (readers may compare it with the elaborate arrangement on F plasmid of *E. coli*). This simplicity has been explained by the lack of an extensive outer membrane and non-requirement of sex pilus for plasmid exchange. Many described plasmids of Gram-positive bacteria are promiscuous. The plasmid pAM 1, isolated from *Enterococcus faecalis* can be transferred to *Staphylococcus aureus* and several other Gram-positive bacteria.

In *E. faecalis* (*E. hirae),* for example, the conjugation is not mediated by sex pili but involves clumping between plasmid-carrying and plasmid-free cells. The former behaves like a typical donor and the latter as the recipient cell. In this species, chromosome several conjugative plasmids have been described, which on the basis of their transferability have been grouped in two categories. Group I, such as pAD1, pOB1, pPD1, pJH2, pAM1, pAM2, and pAM3, transfer at a relatively high frequency in broth (10^{-3}–10^{-1}\donor). Group II plasmids such as pAC1, pIIP501, and p5M15346, on the other hand, transfer poorly in broth but efficiently in filter matings. Conjugation is based on the production of a sex pheromone that helps in generating cell-to-cell contact. The pheromone is a small peptide ($\sim$7–8 amino acids long) which is produced and excreted by the recipient and that induces certain donor cells to adhere and facilitate the formation of donor-recipient mating aggregates.

A model has been proposed to explain conjugation in these bacteria. According to this, a plasmid-free recipient produces, for example, two different chromosomally encoded pheromones, cA and cB. Each pheromone is specific for a donor cell type containing the corresponding plasmid, i.e., cA will stimulate conjugation with a donor carrying the plasmid pA, and cB with the donor having the plasmid pB. These two isogenic donor strains must also be present in the vicinity so as to successfully initiate the process. All three strains produce binding substance (BS) from a chromosomally located gene. The pheromone, cA or cB, then induces the synthesis of aggregation substance (AS) by the donor cell. For this, the pheromone diffuses into the donor cell and reacts with a plasmid gene-encoded product called as responding substance. Again, two types of these substances, e.g., RcA and RcB, are produced to match the pheromone and induces the synthesis of AS. Aggregation substance is a proteinaceous microfibrillar material that is deposited on the cell surfaces and recognizes BS on the potential recipient cells. The interaction between AS and BS causes aggregation between donor and recipient and promotes conjugal transfer. The donor plasmid also possesses an inhibitor gene (icA and icB, in this example) that suppresses the chromosomal pheromone gene in the donor and thus prevents the endogenous production of pheromone. In other words, donor cannot "self-stimulate" themselves under non-conjugating conditions. The pheromone also triggers the plasmid transfer to complete the conjugation process. After receiving the plasmid, the recipient cell ceases to produce the particular hormone and the whole circuit is broken once the two cells separate and the synthesis of AS stops (Fig. 3.19).

The pheromone-mediated plasmid conjugation system is highly efficient. Many Gram-positive bacteria, however, possess a non-pheromone-based, low efficiency plasmid conjugation system. The examples are *Streptococcus pneumoniae* and *S. sanguis*.

Another unique system available in some Gram-positives is the transfer of elements called conjugative transposons. The intercellular transposition of such transposons such as, Tn916, Tn918, Tn920, and Tn1545 is resistant to deoxyribonuclease and requires cell-to-cell contact. They also possess Tra functions to promote their transfer. These transposons code for two proteins, Xis-Tn and Int-Tn, which like phage lambda's Xis and Int promote site-specific integration of DNA with minimum of homology requirement. The excision of the element is initiated by a six base pair staggered cut, known as coupling sequence. The excised element circularizes through base pairing between these coupling sequences but the ends are not complementary. Unlike the conventional

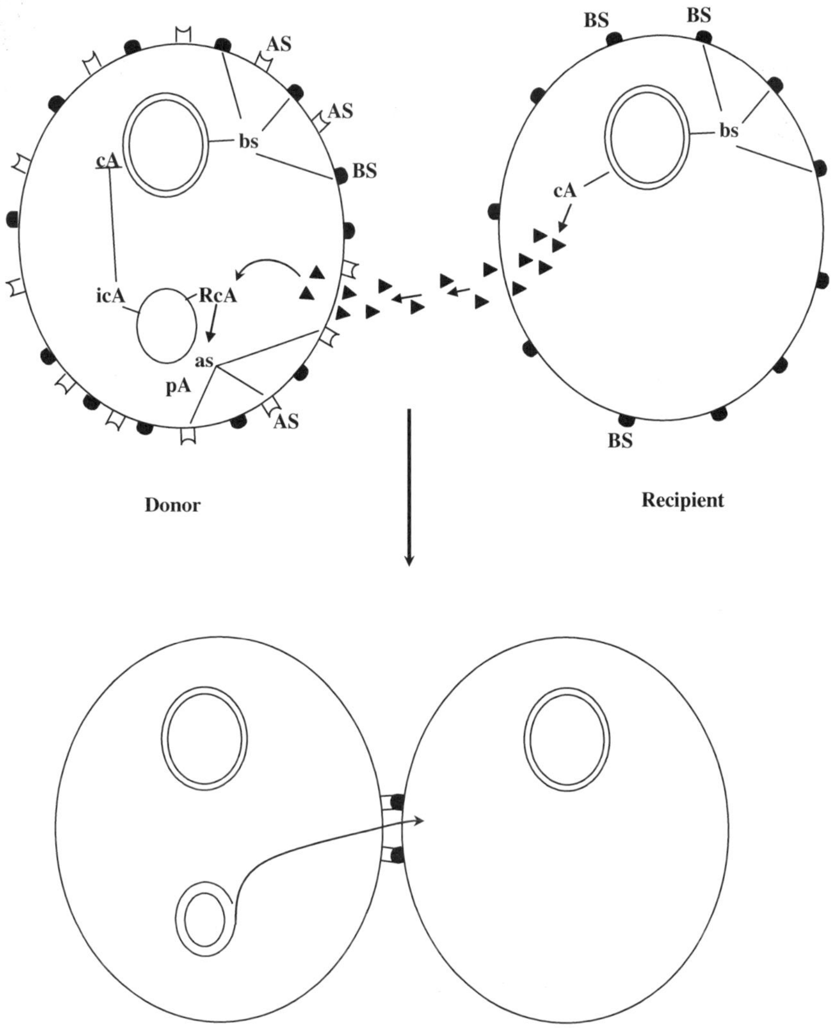

Fig. 3.19 Plasmid-specific pheromone-based conjugation system in *Enterococcus hirae*. *pA* plasmid, *cA* chromosome coded pheromone, *BS* binding substance, *AS* aggregation substance, *icA* inhibitor of cA, *RcA* responder of cA

transposons, Tn916 and its relatives are known to be promiscuous allowing their transfer to many Gram-positive bacteria.

In *Bacillus subtilis*, the chromosomal DNA is transmitted without the intervention of an autonomous plasmid. This transfer requires cell-to-cell contact, initiates at a single site (origin of transfer), and correlates with chromosomal replication, as in Hfr-mediated transfer in *E. coli*. A possible difference in the two systems is that in *B. subtilis* such a transfer is limited to the first replication of the chromosome of a germinating spore.

In mycoplasma, *Spirosoma citri*, chromosomal transmission system resembles that of *B. subtilis*. Consistent with the presumed lineage of mycoplasma from Gram-positive bacteria, conjugation has been demonstrated between *Mycoplasma hominis* and *E. faecalis*.

3.12.1 Actinomycetes

The interest in this group emanates from the fact that several of its members are antibiotic producers. A conjugation mechanism occurring in certain members of actinomycetes differ markedly from other Gram-positive bacteria. One of the best-studied members is *Streptomyces coelicolor*, where crosses can be effected by growing together two genetically different strains on non-selective media and then transferring the resulting cell mass onto the selective plates. This allows the identification of recombinant types of progeny.

The first conjugative plasmid from a Gram-positive organism was described from *S. coelicolor* A3. The plasmid SCP1, shown to influence the exchange of chromosomal genes, is a linear double-stranded DNA plasmid of 350 kb. Several comparisons can be made between SCP1 and *E. coli* F factor. The cell lacking SCP1 ($SCP1^-$), earlier called ultrafertility type or (UF) are like F^- *E. coli*. However, crosses between two such strains do result in low level recombinants for chromosomal markers in *S. coelicolor* but not in *S. lividans*. SCP1 strain (IF or initial fertility) behaves as F^+, as the transfer of the plasmid is much more common than chromosomal markers. A cross between $SCP1^+$ to $SCP1^-$ show preferential inheritance of donor chromosomal markers suggesting an Hfr-like situation. A number of Hfr-like derivatives of SCP1 initially called NF or normal fertility have the plasmid integrated by a recombination between a copy of IS446 on the plasmid and one or two copies on the chromosome. Transfer of chromosome by such strains appears to be bi-directional and the recombinants of NF to UF cross are all NF. On both the accounts, the SCP system appears to differ from that of Hfrs of *E. coli*. Various SCP-prime plasmids have been isolated in *S. coelicolor*, which can mediate the mobilization of donor chromosomal markers. They can also be integrated generating both stable and unstable Hfr-like strains.

Several lesser known plasmids have also been described in *S. coelicolor*. One such small, self-transmissible plasmid pIJ101, often used as a cloning vector, appears to carry a single *tra* gene. All in all, more needs to be understood about the conjugational genetic exchange in Gram-positive bacteria in comparison to their Gram-negative counterparts.

3.13 Conjugation in Genetic Analysis

Conjugation is one of the principal mechanisms of gene transfer in bacteria. Such a mechanism allows the transfer of a chunk of DNA from a donor cell to a recipient where recombination can take place to generate new combination of alleles or in other words create genetic variability in the population. This method is also used for effecting plasmid transfer or other genetic elements such as transposable elements.

The other important application of this process is to map a gene of interest on the chromosome, as described in detail in this chapter. Such maps may have further implications in the genetic analyses in bacteria. This process also contributes toward the creation of F-primes that have been employed in dominance analysis in otherwise haploid bacterial systems, as well as in understanding of the regulatory aspects.

Further Readings

Babic A, Lindner AB, Vulic M, Stewart EJ, Radman M (2008) Direct visualization of horizontal gene transfer. Science 319:1533–1536

Bachmann BJ (1990) Linkage map of *E. coli* K-12. Microbiol Rev 54:130–197

Berlyn MKB (1998) Linkage map of *E. coli* edition 10: the traditional map. Microbiol Mol Biol Rev 62(3):814–984

Clark AJ, Adelberg EA (1962) Bacterial conjugation. Annu Rev Micrbiol 16:289–319

Frost LS, Koraimann G (2010) Regulation of bacterial conjugation: balancing opportunity with adversity. Future Microbiol 5(7):1057–1071

Frost LS, Ippen-Ihler K, Skurray RA (1994) Analysis of the sequence and the gene products of the transfer region of the F sex factor. Microbiol Rev 58(2):162–210

Grohman E, Sadowy E, Hryniewicz W (2003) Conjugative plasmid transfer in Gram-positive bacteria. Microbiol Mol Biol Rev 67(2):277–301

Hayes W (1953) Observations on a transmissible agent determining sexual differentiation in *Bacterium coli*. J Gen Microbiol 8:72–88

Jayaraman R (2011) William Hayes and his Pallanza bomb shell. Resonance 13:911–921

Kohiyama M, Hiraga S, Matic I, Radman M (2003) Bacterial sex: playing voyeurs 50 years later. Science 301:802–803

Lawley TD, Klimke WA, Gubbins MJ, Frost LS (2003) F factor conjugation is a true type IV secretion system. FEMS Microbiol Lett 224:1–15

Lederberg J, Tatum EL (1946) Gene recombination in *E. coli*. Nature 158:558

Llosa M, de la Cruz F (2005) Bacterial conjugation: a potential tool for genomic engineering. Res Microbiol 156:1–6

Ogar P, Farrand SK (2002) Two opines control conjugal transfer of an *Agrobacterium* plasmid by regulating expression of separate copies of the quorum-sensing activator gene *traR*. J Bacteriol 184(4):1121–1131

Sardesai AA, Gowrishankar J (2008) Joshua Lederberg—a remembrance. J. Genet 87(3):311–313

Smith GR (1988) Homologous recombination in prokaryotes. Microbiol Mol Biol Rev 52(1):1–20

Wardel E, Sadowy E, Hryniewicz W (2010) Complex nature of enterococcal pheromone-responsive plasmids. Pol J Microbiol 59(2):79–87

White CE, Winnas SC (2007) Cell–cell communication in the plant pathogen *Agrobacterium tumefaciens*. Phil R Soc B 362:1135–1148

Williams SL, Schildebach JF (2007) TraY and integration host factor *oriT* binding site and F conjugal transfer: sequence variations, but not altered spacing, are tolerated. J Bacteriol 189(10):3813–3823

Wollman E, Jacob F, Hayes W (1956) Conjugation and genetic recombination in *E. coli* K-12. Cold Spring Harb Quant Biol 21:141–148

Wong JJ, Jun L, Edwards RA, Frost LS, Glover JNM (2011) Structural basis of co-operative DNA recognition by the plasmid conjugation factor. TraM Nucleic Acid Res 39(7):1–14

Transformation

4

4.1 Discovery

In 1928, Fred Griffith, a British pathologist was studying the pathogenicity of pneumococcal infections in mice that also causes pneumonia in humans. Based on the disease response, pneumococcal strains could be classified as virulent or disease-causing and avirulent or non-pathogenic. He observed that the virulent strains of the causative organism, *Pneumococcus* (*Streptococcus*) *pneumoniae* can be easily distinguished from the avirulent ones by the fact that they produced smooth glistening colonies on agar surfaces and thus were referred to as S strains. Several such S strains could be identified based on a polysaccharide capsule that carries antigenic properties. Such strains were labeled as SI, SII, SIII, etc., basically differing in the composition of their capsular polysaccharides, and thus, eliciting different antigenic responses.

Mutants could be obtained from these S strains and have been labeled as RI, RII, RIII, etc., respectively. These mutants produced rough colonies and were avirulent in their disease response. Reversion from R to S is very rare but is always true to its type, i.e., RI to SI, RII to SII, and so on. Griffith was also able to show that virulent S strain can be attenuated by heating the cell suspension to ~80 °C. Thus, heat-killed (hk) S strains and any of the R strains are non-pathogenic when injected into the peritoneal cavity of mouse. However, when the injected material consisted of a mixture of heat-killed S type (SIII, for example) and a live R type (RII, for example), the disease appeared and mice died of pneumonia. From the bloodstream of these dead mice, he could isolate live SIII bacteria (Fig. 4.1).

Griffith termed this phenomenon transformation, wherein a transforming principle released by the heat-killed cells of S type bacteria transformed the live R type counterpart. The transformation represented a true genetic change as the S type of bacteria so formed continued to produce the same type of cells after subsequent cell divisions. Transformation was in fact the first method of gene transfer discovered in bacteria. This historical reason, as well as its contribution toward revealing the chemical nature of the gene, and also as a process to deliver and express foreign genes in a suitable host, makes this phenomenon of central importance in Biology. Here, we shall discuss the process of transformation as a method of gene transfer in bacteria.

Soon after the work of Griffith, efforts were directed to devise an in vitro transformation protocol, so as to bring this process under straightforward scientific parameters. Dawson and Sia (1931) were able to convert live R type bacteria into S type when incubated with heat-killed S type bacteria in a test tube. This could be further substantiated by inclusion of anti-R-antisera in the reaction mixture that will coagulate the live R type bacteria, in case transformation did not work. Alloway (1933) brought about another refinement by using a cell-free filtered extract from the S type bacteria and yet

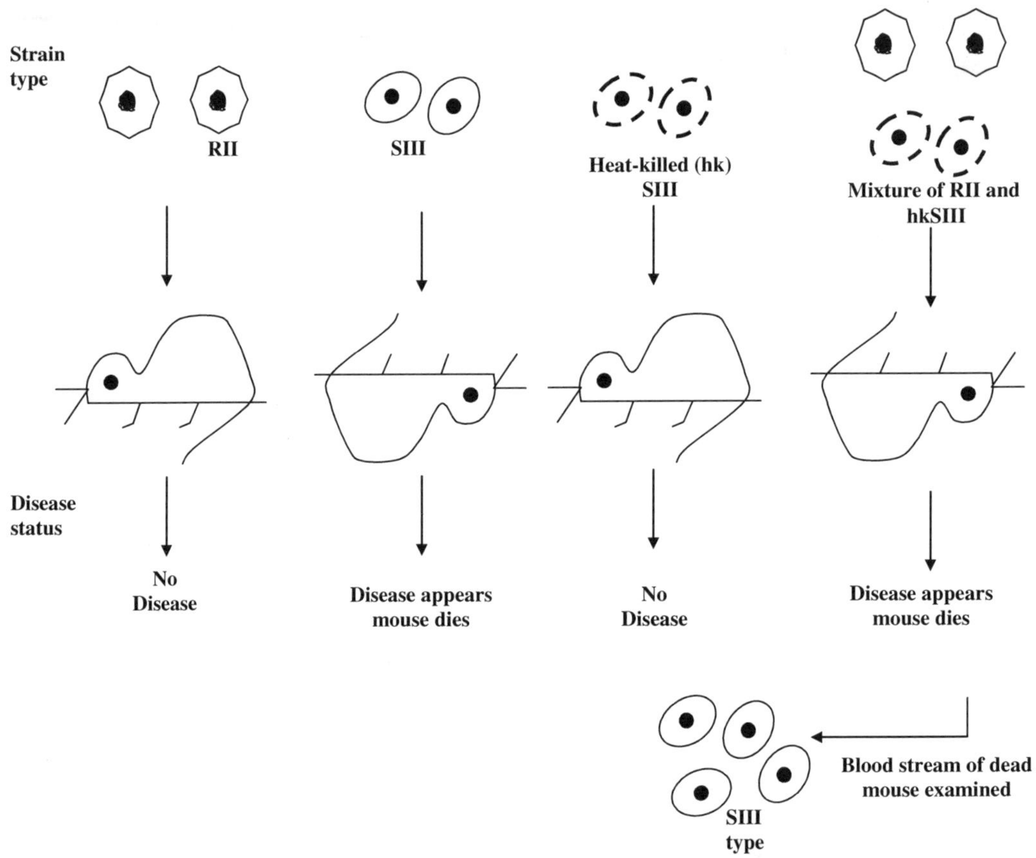

Fig. 4.1 Griffith's experiment to demonstrate transformation in *Pneumococcus* (*Streptococcus*) (RII → SIII)

demonstrating the transformation of an R type bacterial strain.

4.2 The Nature of Transforming Principle

The nature of transforming principle, as predicted by Griffith, was a matter of great speculation for a long time. It was suggested that acquisition of capsular material released by the killed bacteria by non-encapsulated strain may lead to transformation. However, such a proposition could be ruled out by the fact that transformation brings about a permanent genetic change such that the new phenotype is perpetuated generation after generation. It was at Rockefeller Institute, USA that conclusive experiments were conducted by Avery et al. (1944) to reveal the nature of the transforming principle. From a series of experiments, involving different hydrolyzing enzymes, and purification of some macromolecules, they concluded that transforming principle is nothing but deoxyribose nucleic acid (DNA). This information was important as proteins were considered as strong contenders of being identified as the genetic material. They demonstrated that treatment of cell-free extract with protein digesting enzymes, trypsin or chymotrypsin, did not obliterate its transforming ability. Similar was the effect of ribonuclease treatment, ruling out RNA, but treatment with DNase completely abolished this process (Fig. 4.2).

The implication of DNA as being the active agent in transforming principle was significant for two reasons. First, this molecule was studied more extensively as the chemical form of genetic

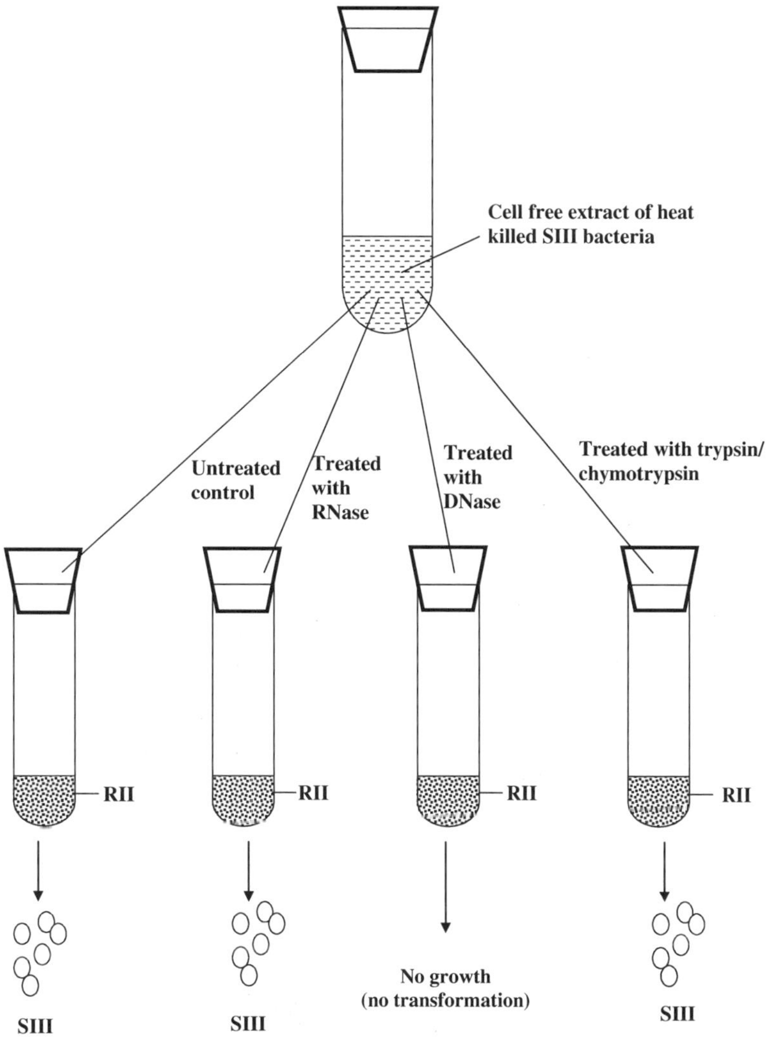

Fig. 4.2 Diagrammatic representation of Avery, Macleod, and McCarty's experiment to demonstrate that DNA constitutes the transforming principle in Pneumococci/Streptococci

material, and several cellular functions were explained on the basis of its structure and behavior. Second, its genetic connotation became established when bacteria other than streptococci were shown to undergo transformation. In 1951, Alexander and Leidy demonstrated the conversion of R type into type specific S strain by the DNA preparation derived from the latter in *Haemophilus influenzae*. In the same year, Ephrussi-Taylor showed that capsular phenotypes other than R and S could also be transferred in pneumococci. The biggest boost, however, came through the work of Hotchkiss (1955) who demonstrated that penicillin resistance could be transferred to sensitive cells through the DNA preparation derived from the resistant cells. This provided an easy way to select and identify transformants, and opened the avenues for extensive investigations on the process *per se*.

4.3 Transformation as a Method of Gene Transfer

Transformation as a parasexual method of gene transfer has now been described for a large number of bacteria, but the three model systems remain to be *Streptococcus pneumoniae*, *Haemophilus influenzae*, and *Bacillus subtilis*. In addition, many bacteria like *E. coli* could be persuaded to pick-up exogenous DNA because transformation has become a critical step in recombinant DNA technology. These studies have led to the classification of this process as (i) natural transformation and (ii) artificial transformation.

The one common protocol for transformation is to treat the appropriate recipient bacteria in a particular state (competence) with the DNA derived from a donor cell. The two are incubated for a specific time and then in some cases a DNase wash is given to get rid of extraneous DNA. The recipient cells are grown for a specified time (one growth cycle), and then plated on a selection medium and incubated till transformants appear. Essentially, transformation consists of the following steps:

(i) release of donor DNA, naturally or experimentally;
(ii) preparation of a recipient culture competent to interact with donor DNA;
(iii) uptake of donor DNA, and its associated processing, if required; and
(iv) integration of donor DNA at homologous site in the recipient chromosome. In plasmid transformation, integration is not required.

4.4 Natural Transformation

As described above, many bacteria can undergo transformation naturally. The three model systems mentioned so far, viz., *S. pneumoniae*, *H. influenzae*, and *B. subtilis* have provided most of the information on mechanistic details, though the list of bacteria that are naturally transformable is ever growing.

4.4.1 Release of Donor DNA

All naturally transforming systems are known to release DNA that can be utilized as a transforming agent. It was initially thought that donor DNA is released from the dying cells and could be utilized by the appropriate recipient cells for any transformation to occur. It is, however, now clear that this event takes place in the same population that is undergoing competence development (see below) by a coordinated set of steps. While the majority of cells induced to competence get prepared to receive the donor DNA, a subset of cells are lysed to release the donor DNA. This mechanism appears to have a lot of implications under the natural environment. Most importantly, the naturally competent streptococci, for example, can avoid taking up too much DNA from unrelated bacteria, lest its mismatch repair system gets overloaded and leads to large-scale damage to the recipient's genome. Streptococci have perhaps evolved this way, a mechanism to distinguish self- from non-self.

DNA uptake and DNA release, are both induced by the same signal transduction pathway (*comCDE*) that also regulates the competence development in *S. pneumoniae*, as described below. The proteins involved in DNA release are the members of *comE* or *comX* regulon. DNA release is dependent on CBP (choline binding peptide), many different types of which are synthesized by pneumococci. Three murein hydrolases, LytA, LytB, LytC, which are anchored to the cell wall through their choline binding domains, have been implicated in this step. Thus competence induction triggers autolytic amidase, LytA, which is a member of the *comX* regulon, and the autolytic lysozyme, LytC. Although LytA is expressed in non-competent cells also, it gets significantly upregulated during competence development. The net effect is the controlled lysis of the cell and the release of the DNA. These enzymes are under strong negative regulation so as to avoid uncontrolled digestion of the cell wall otherwise. In streptococci, ~5–20 % of the cells in a competent population lyse and act as donors of DNA, though this will also depend upon the growth conditions. This coordination between competence development and release of donor DNA ensures that the two critical requirements for the process are available at the same time.

4.4.2 Competence

One early observation was that the recipient cells cannot be transformed at all times. In other words, they needed to attain a state and status that came to be called as competence, to be able to get transformed. It has thus become a common practice to prepare the competent cells before any transformation is planned. In fact, the difference between the natural and artificial systems is based on how these cells become competent. While in the former case, this happens during normal growth, in the latter, competence needs to be induced.

If we follow the emergence of competence in terms of growth profile, all the three systems follow a typical pattern. The competence arises during late exponential\early stationary phase and remains for a short time in streptococci and *Haemophilus,* but in *B. subtilis* it may stay for several hours. This can be estimated either by determining the number of transformants/ml or by transforming with the phage DNA and counting the PFU/ml, over the growth period. As seen in these bacteria, transformation frequency increases with the bacterial growth, reaching maximum in the late exponential\early stationary phase cells (Fig. 4.3).

One can also determine the proportion of competent cells in a population. Genetically, this can be determined by estimating the observed double transformation frequency of two unlinked markers and calculating the expected double transformation frequency as a product of two single transformation frequencies of the same genes. The ratio of expected to observed double transformation frequencies gives the proportion of competent cells in the population. When estimated, this is high (100 %) in *S. pneumoniae* and *H. influenzae* but low in *B. subtilis* (15–20 %). Similar information is obtained by providing labeled donor DNA and tracing its uptake by autoradiography. This suggests that while in *Streptococcus* and *Haemophilus* almost all the cells become competent, in *B. subtilis* only ~10–15 % cells acquire competence and thus are capable of participating in transformation.

Another important characteristic of competence in streptococci is that it is transferable between the cells. Competence may, therefore, initiate in a small group of cells but spread rapidly through the population. Mixing a small number of competent bacteria with a large number of non-competent ones can bring about competence in the latter. Similarly, two such cell populations separated by a Millipore filter do not hamper the transfer but a dialyzing membrane placed to separate the two populations of cells can completely stop it. This led to the suggestion that some macromolecule is involved, the proteinaceous nature of which was subsequently

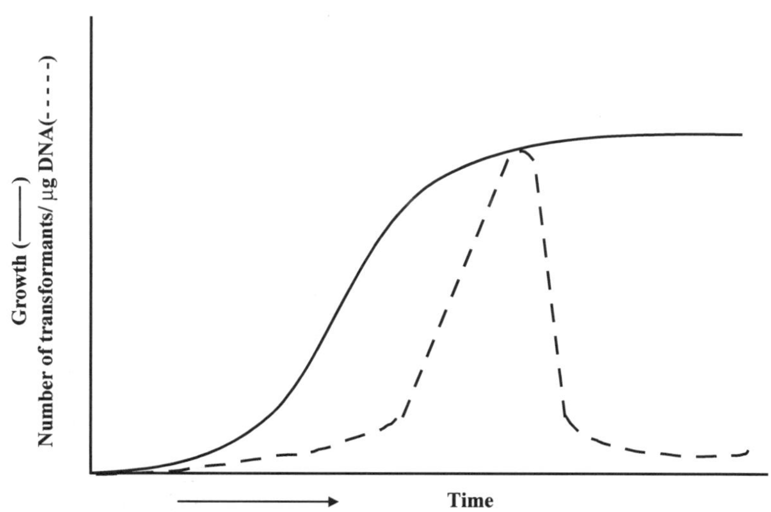

Fig. 4.3 Growth phase-dependent competence development as assessed by the number of transformants/µg DNA (- - - - - -) during natural transformation in streptococci

confirmed when the activity of the isolated factor was shown to be sensitive to trypsin. The factor was called as competence factor. Inhibition of RNA and protein synthesis, but not DNA synthesis affected the acquisition of competence.

The role of the cell wall in streptococcal transformation was demonstrated when the cell wall deprived of choline resulted in the cell's failure to respond to competence factor. It is also known that autolytic enzymes degrade the cell wall only if it contains choline. The new choline molecules are deposited onto the cell surface at a growing zone located at the equatorial region of the cell. This tends to suggest that DNA during transformation may enter the cell through this growth zone. The content of autolytic enzymes also showed a good correlation with the emergence of competence. Streptococci are able to attain competence only if incubated in a specific medium suggesting that some changes in the cell envelope during incubation may facilitate the DNA's accessibility to receptor sites. The process of competence development is a highly regulated and complex phenomenon involving many molecular events that have now been studied and revealed.

As described above, the development of competence in streptococci is based on the release of a protein competence factor, also known as competence stimulating peptide (CSP). It is a 17 amino acids long peptide synthesized in a cell density-dependent manner. This quorum sensing-based system allows a gradual build-up of CSP till it reaches a threshold concentration that leads to rapid spread of competence through the pneumococcal population. CSP is a product of *comCDE* operon, wherein *comC* codes for a 41 residue long Pre-CSP from which the final product is cleaved out. Another operon consisting of *comA* and *comB* provides the processing and export function by an ABC transporter system. In addition, a two-component regulatory system is also pressed in action. As the CSP concentration builds up, it is sensed by a membrane located histidine kinase (sensor), a product of *comD*. This involves the phosphorylation of ComD, which in turn then passes on the phosphate

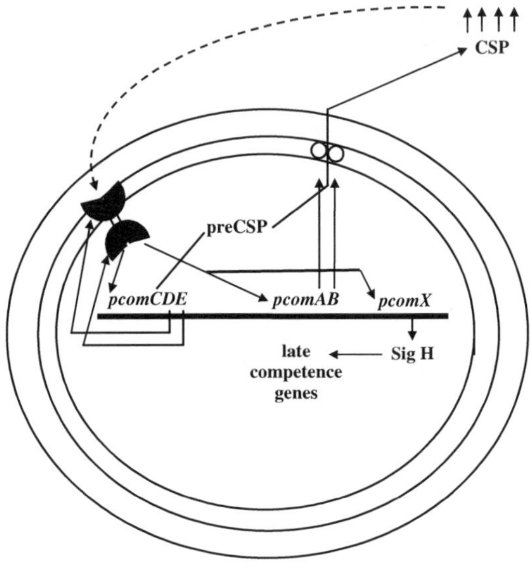

Fig. 4.4 Competence development in streptococci. It involves both quorum sensing and 2-component regulation

group to ComE (response regulator).The phosphorylated ComE, in turn, autoinduces the expression of both *comAB* and *comCDE* operons. It has also been postulated that ComE induces the production of a secondary competence-specific sigma factor, the *comX* product, which then allows transcription from the late competence genes (Fig. 4.4).

ComE may also play a dual role in regulating competence development in a CSP concentration-dependent manner. The product of *sigH* gene has been implicated in controlling the duration of competence and post-competence development. Streptococci like bacilli (see below) also undergo differentiation, such as reduced growth, longer chains of cells, autolysin activation, but most importantly 16 different proteins associated with competence also appear. The operons coding for some of these proteins also share a common consensus sequence, TACGAATA, known as "*com* box" in the promoter. CSP-based competence response, often equated with a pheromone, functions in a highly strain-specific manner, which in fact stems from strain specific allele of *comD*. Different *comD* alleles reflect different pheromone specificity (pherotype). The pherotype diversity is perhaps

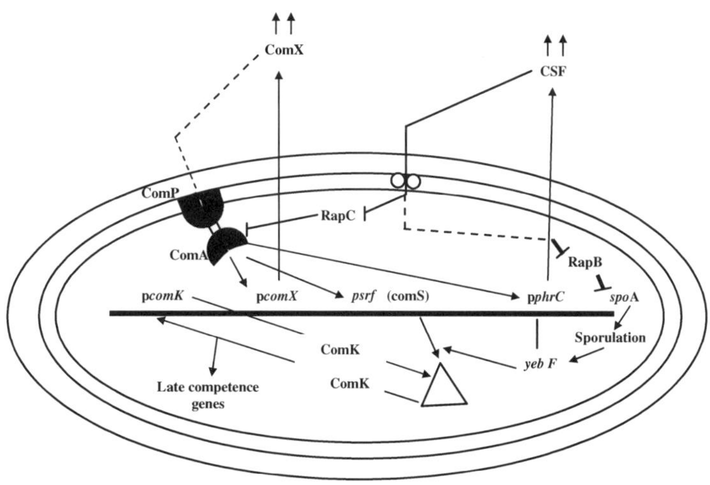

Fig. 4.5 Competence development network in *Bacillus subtilis*. Like pneumococci, it involves quorum sensing and two-component regulation

evolved to ensure that a bacterium does not develop competence when surrounded by more distantly related bacteria.

Bacilli represented by *B. subtilis* show a number of differences from the streptococcal system. First, *B. subtilis* can achieve competence in defined medium but only ~20 % cells become competent. These cells are smaller in size and metabolically inactive. For example, the competent cells are resistant to ampicillin, suggesting that these are non-dividing. Similarly, competent thymine-requiring mutant cells do not undergo thymine-less death and produce transformants, without any appreciable loss, in a medium lacking thymine. One can infer that these were not synthesizing DNA.

Like streptococci, competence development in *B. subtilis* has also been worked out in detail. Once again, this involves both quorum sensing and two-component regulatory system (Fig. 4.5). Bacilli undergo sporulation while the cells are attaining competence. Thus further interlinks are expected. During competence development, two peptides are exported out of the cells, both of which function through a quorum sensing system. One ComX, a competence pheromone is made up of 9–10 amino acids (modified), and the other is a competence stimulating or competence and sporulation factor (CSF). The ComX, is a 55 amino acids peptide, and the competence pheromone is generated by a cleavage at C-terminus, releasing a 10 amino acids product. It binds to the membrane located sensory protein, ComP (a histidine kinase), which in the process gets auto-phosphorylated, and the phosphorylated ComP then passes on the phosphate group to ComA, a response regulator or responder. When *Bacillus* cells are undergoing competence, depletion of nutrients and other signals invoke the sporulation pathway. This pathway bypasses ComX and ComP, but feeds into ComA and works by regulating the cytoplasmic levels of CSF. The CSF is synthesized from *phrC* gene and externalized CSF is imported back by an oligopeptide permease that is a part of the sporulation pathway. The CSF also functions by inhibiting a protein, RapC that acts as a phosphatase for ComA. The more the RapC is inhibited the more phosphorylated ComA will be available to induce competence. In other words, the import of CSF modulates the dephosphorylation of ComA.

The phosphorylated ComA induces *srf* operon, coding for the lipopeptide antibiotic, surfactin. One important component of *srf* operon is *comS*, which is also activated by ComA. Thus, the dephosphorylation of ComA prevents activation of *comS*. ComS functions by releasing ComK from a complex consisting of ComK, ClpC, MecA, and ClpP, formed during the log phase of growth when bacilli are not competent. The free ComK that acts as the master transcription regulator then activates the transcription of all the necessary competence

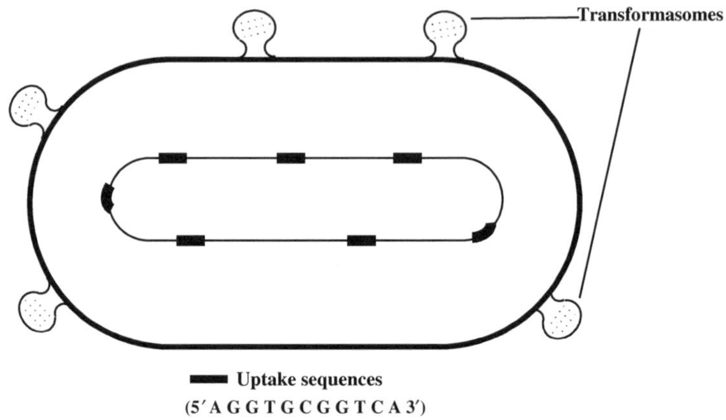

Fig. 4.6 *Haemophilus influenzae* competent cells develop membranous vesicles, transformasomes. This bacterium carries 11 bp sequence called uptake sequences that the donor DNA must possess for its uptake

proteins (at least 13; ComGA, -GB, -GC, -GD, -GE, -GF, -GG, -EA, -EB,-EC, -FA,-FB, -FC) as well as ~100 other proteins, some of them involved in DNA repair and recombination (Fig. 4.5). ComK is turned off after about 2–3 h post induction by being targeted for proteolysis. The protective action of ComS is itself helped by *ylbF* gene product that appears to be a part of sporulation pathway. Once developed, competence stays for a long time in bacilli. *B. subtilis,* competence and sporulation factor (CSF), is a pentapeptide and is also controlled by quorum sensing. The five amino acids at the C-terminus of the 40 amino acid precursor, PhrC are cleaved out to form the CSF molecule. CSF accumulates extracellularly as a function of increasing cell density. However, the signaling role of CSF is intracellular. Extracellular CSF is imported into *B. subtilis* cells by an ABC-type oligopeptide transporter called OPP. When intracellular concentration of CSF is low, it binds to and inhibits a ComA-specific phosphatase, RapC, thus increasing the levels of phosphorylated ComA. As described earlier, the phospho-ComA is the response regulator, controlling the expression of genes required for competence (see Fig. 4.5).

High intracellular levels of CSF, on the other hand, inhibit competence and induce sporulation. At high concentration, CSF inhibits the expression of *comS* that results in increased proteolysis of ComK, the protein required for the decision to commit competence. The mechanism by which CSF stimulates sporulation is analogous to that for competence. During sporulation, CSF inhibits another phosphatase, RapB. RapB dephosphorylates a response regulator, SpoA, involved in promoting sporulation. High levels of phospho-SpoA favor a switch in commitment from competence to sporulation pathway. Presumably, adjustment of the concentration of CSF above or below the critical level tips the balance in favor of one lifestyle over the other, allowing *B. subtilis* to correctly choose and commit one of the two very different terminal fates. Many bacilli can attain competence in the natural environment. Both Gram-positive systems are comparable as far as the cell wall modifications and the controlled release of autolysins are concerned.

Gram-negative *Haemophilus influenzae* and *Neisseria* are naturally transformable. *H. influenzae* attains low-level competence during late exponential phase. The competence level can be greatly improved if the cells are shifted from a defined growth medium to non-growth medium. Transfer to rich medium destroys competence. Competence is quite stable and competent cells develop many membranous vesicles called transformasomes. These transformasomes are seen budding from the outer membrane and contain proteins of varying sizes (Fig. 4.6). DNA uptake in *H. influenzae* is highly specific and is dependent on the recognition of a conserved sequence on the donor DNA. The proteins carried by transformasomes have been implicated in this recognition reaction as noncompetent cells carried very few vesicles and lacked the concerned proteins.

The role of cAMP has also been documented in *Haemophilus* competence. A regulator gene, *sxy (tfoX)*, has been identified through the isolation of two types of mutations; one type makes the cells non-competent and the other results in constitutive competence. This gene is also required for the induction of late transformation genes, such as *dprABC* and *comF*. The genes *dprA, comA-F,* and *pilA-D* are flanked by an upstream 26 bp palindromic site at which the regulatory elements may act. *Haemophilus, Bacillus*, and *Streptococcus* all carry orthologues of type IV pili components, important for acquiring competence, but these strains do not produce pili. On the other hand, in *Neisseria, Pseudomonas, Moraxella*, and *Legionella* fully formed pili are essential for transformability.

4.4.3 Cell–DNA Association

The next step in transformation is the association of a competent recipient cell with appropriate donor DNA. This association involves binding of donor DNA to the cell, in a two-step process (Fig. 4.7). First, the DNA is reversibly bound (loose binding), followed quickly by an irreversible or tight binding. In the latter state, DNA is said to be committed for uptake. In the former state, DNA may remain sensitive to shearing and DNase but cannot be removed by washing.

In *S. pneumoniae*, the shift from loose to tight binding takes place within a few seconds and then binding continues for about 30 min. It has been estimated that each pneumococcal cell can bind anything between 30 and 80 molecules of DNA. Once tightly bound non-specific carrier DNA can also not compete for binding. As discussed above, during competence development, receptor sites may become accessible to DNA facilitating the binding.

Pneumococcal mutants lacking major endonuclease activity (*endA* mutants) are affected in transformation as well. Mutants having more than 1 % of endonuclease activity are normally transformable, but if they have less than this level, DNA binds but is not processed further. If the bound DNA is re-extracted, following alkali treatment, single-strand breaks seem to have been introduced, approximately 6.0 kb apart. These single-strand nicks are later extended into double-strand breaks by the same enzyme coded

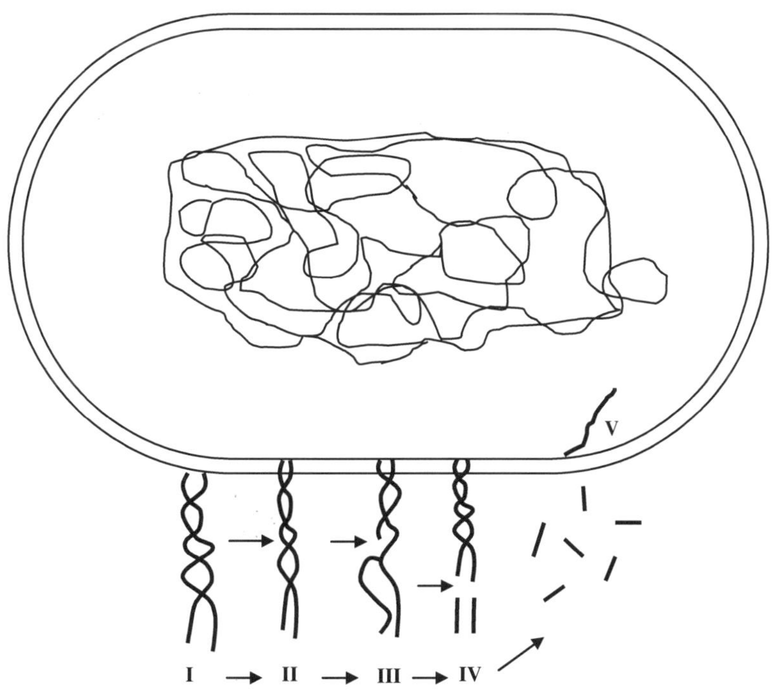

Fig. 4.7 Sequential steps in the association, processing, and uptake of donor DNA during transformation in Pneumococci. I—reversible binding; II—irreversible binding; III—introduction of single-stranded nick; IV—conversion of nick into double stranded break; and V—uptake of single-stranded DNA and the degradation of the other strand. *Note* Several DNA molecules may get bound, processed, and taken up by a competent cell

by *endA*. This processing, therefore, constitutes an important step in DNA uptake (Fig. 4.7).

B. subtilis competent cells possess 30–50 DNA binding sites. DNA binding is rapid and soon the DNA enters DNase-resistant state. Like in streptococci, DNA is fragmented with an average size of 13.5–18 kb. A surface located nuclease (NucA) interacts with the bound DNA. In *B. subtilis,* binding and processing of DNA requires a whole panoply of proteins. The principal one is ComEA. This membrane spanning protein not only binds to donor DNA but may also function in its transport. In addition, a ComG protein is also required but more for providing access to incoming DNA to ComEA. We shall discuss about these proteins in the following section.

In *H. influenzae* as well as *Neisseria*, the binding and uptake requires the specific uptake sequence, called uptake signal sequence (USS) or DNA uptake sequence (DUS). As foreign DNA is unlikely to have such a sequence, these bacteria are able to differentiate between homologous and non-homologous DNA. This 9 bp sequence, 5′-AAGTGCGGT-3′ in *Haemophilus* is known to occur in multiple copies, distributed all over the genome. In *Neisseria*, the USS is a 10 bp sequence, 5′ GCCGTCT GGAA3′. The binding reaction is similar to that of Gram-positive bacteria. Only 2–4 donor molecules are estimated to interact per competent cell but the association is very productive. Unlike Gram-positive systems, EDTA does not affect binding or uptake, suggesting that surface-bound nucleases are not involved. Although the DNA is taken up fast, it first gets associated with membranous vesicles or transformasomes by interacting with surface proteins. Sxy protein has been implicated in the regulation of DNA binding.

All naturally transformable systems share proteins found in type IV pili but many of them do not synthesize pili. Many orthologues of this class of proteins may be involved in constituting the uptake channels. In all the systems, double-stranded DNA is the most preferred substrate. Uptake of single-stranded DNA, RNA, or DNA–RNA hybrids is highly inefficient. In Gram-positive systems, low level transformation is possible with single-stranded DNA but only under harsh, non-physiological conditions, such as, low pH and the presence of EDTA.

4.4.4 Uptake of Donor DNA

As described earlier, the acquisition of DNase resistance is the first step toward DNA uptake. In both streptococci and bacilli, DNA is processed by double-stranded breaks introduced by membrane-bound major endonuclease. In *B. subtilis*, divalent cations such as Ca and Mg are essential for competence development and Mg activates the endonuclease activity. EDTA blocks DNA uptake by chelating these ions. In *S. pneumoniae*, DNA fragment of 6.0–9.0 kb, and in *B. subtilis* of 13.5–18 kb are produced for uptake.

Soon after the binding of DNA, in *S. pneumoniae*, the double-stranded DNA is converted into single strands. While the strand with 3′–5′ polarity is taken up, the other strand is degraded. This appears to be the only change brought about in DNA during uptake as the DNA that enters the cell and the TCA-soluble oligonucleotides released outside account for all the DNA that is added. Some of these steps have been depicted in Fig. 4.7. The single strands then associate with SSB protein so that they are not only protected, but may also help in the subsequent step of integration by RecA-mediated homologous recombination. This is also the stage where homologous DNA is differentiated from the non-homologous ones, even when the latter is taken up. Such a DNA, therefore, is not integrated and is soon degraded. Large numbers of proteins participate in the formation of the aperture where DNA binds and gains entry into the cell. Many of these are similar to those described in *B. subtilis*. While CelB provides the membrane pore for DNA entry, CelA is involved in DNA binding. The Cgl proteins (CglABCD) are similar to *comG* (ORF1-14) product of bacilli and may also function to modulate the channel or the wall to facilitate DNA entry. The energy requirement for DNA uptake is met by ComF.

In *B. subtilis*, DNA uptake occurs through a three-step mechanism. First, the transfer of DNA through the cell wall is mediated by the products of *comG* operon, constituting a pilus-like (pseudopilus) structure. While ComGC contributes the major pilin-like subunits, ComGA is the cytoplasmic ATPase required for the assembly of the pseudopilus, as well as for repressing the growth and cell division during competence phase. The pseudopilus is proposed to be a dynamic structure that may extend and retract so as to pull the extracellular DNA for uptake across the cell wall. In the second step, double-stranded DNA gets bound to membrane located ComEA. This is followed by the cleavage of the DNA into shorter fragments by NucA endonuclease and subsequent conversion into single-stranded DNA. The final step involves the uptake of these single strands into the cytosol through the ComEC membrane permease. ComEC is proposed to form a single-stranded DNA-specific membrane channel, and ComFA is an ATPase that may energize the uptake of DNA. The net result is that only one DNA strand enters the cytosol and the other is degraded. The polarity of single-stranded DNA taken up can either be 3′–5′ or 5′–3′. This DNA enters into a complex and does not recombine with the recipient genome immediately. Degraded DNA can be detected in the medium suggesting that DNA digestion takes place outside the membrane. Several proteins from bacilli have orthologues in *Streptococcus*, as described above. In bacilli proton motive force is essential for DNA uptake. Recent results have shown that the whole competence machinery is located at the cell pole.

Both *Haemophilus* and *Neisseria* differ from other systems in having a requirement for specific recognition sequence in the donor DNA. Only such DNA can associate with the recipient cell, in fact the binding reaction takes place in the transformasomes than directly in the cell . The competent cells of *Haemophilus* have many more transformasomes than the non-competent ones. The mutant *com*-51 that lacks competence, in fact, sheds these vesicles in the medium. Although the binding sites are limited, the uptake is so rapid that transformasomes have been implicated in the transport and delivery of the donor DNA in the vicinity of recipient chromosome.

Soon after uptake, the DNA is converted into single-stranded form but since such single strands are not detected in the cytoplasm they may be equally rapidly taken up. In other words, the conversion of double strands into single strands may be a part of the integration of the donor DNA into recipient genome. Mutants of *rec2* gene are defective in transport and the DNA remains trapped in the transformasome. Rec2 is a homologue of ComEC and this suggests that Rec2 also forms a channel. Com101 mutants on the other hand neither degrade DNA into single strands nor translocate them. Functionally, they appear similar to ComF of *B. subtilis*. Both Rec2 and Com101 are regulated perhaps by Sxy protein. Several other functions essential for transformation consist of directing the DNA to appropriate receptors, conforming competence proteins for DNA uptake, translocation of DNA across the cell membrane, activation of autolytic enzymes enabling DNA passage through the peptidoglycan layer, etc.

Roughly one uptake recognition site per 4.0 kb length of DNA has been estimated in *Haemophilus*. Statistically, this is much higher than would be expected by chance. The evolutionary significance of specificity and maintenance of these uptake sequence is not very clear. One could only presume that it helps the strain to maintain its species identity. EDTA does not prevent DNA uptake, suggesting that external nucleases are not involved in this process.

4.4.5 Integration of DNA

The final and a very crucial step in natural transformation is the integration of taken up DNA to yield the transformants. In all the three systems described above, the DNA for this step is in a single-stranded state.

Using heavy density-labeled donor DNA, the size of integrated DNA has been estimated to be ~5.0–10 kb in *S. pneumoniae*, ~8.5 kb in

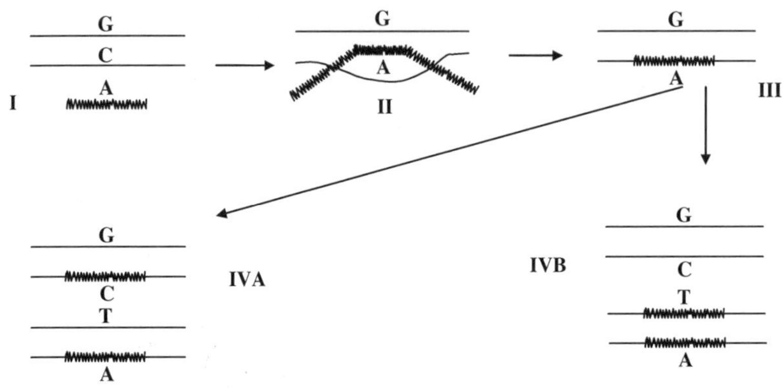

Fig. 4.8 Possible mechanism of generating transformants. I—Single-stranded donor DNA pairs with the homologous sequence in the recipient; II—donor DNA is assimilated and pairs with the complementary sequence; III—all extra DNA is removed and donor DNA is covalently attached to the rest of the recipient DNA, generating a heteroduplex; IV—the heteroduplex is resolved either by IVA: two-way mismatch correction or IVB, by replication. Both will lead to one non-transformed and one transformed event. Single base difference is depicted to differentiate donor and recipient gene sequence

B. subtilis, and ~18 kb, in *Haemophilus.* Integration was studied by using *thy*$^-$ (thymine-requiring) mutant strain and growing it to acquire competence in presence of heavy isotope, 5-BU. Such cells were transformed with non-labeled (light) donor DNA. DNA was extracted from the transformants and fractionated based on the density label. All donor activity, as detected by the label, was restricted to the lighter fraction. This suggested that the information of a single strand of donor DNA is sufficient, and during integration one donor strand must be replacing the homologous sequence in the recipient DNA, through a reaction supported by RecA-like function.

Integration is thus a multistep process. After the delivery of donor DNA in the vicinity of the recipient chromosome, an unstable donor–recipient complex is first formed. This is extended to form a stable non-covalent complex to begin with, followed by a covalent complex formation in which the corresponding relevant recipient strand is removed. In essence, these steps appear very similar to Meselson and Radding's D-loop recombination pathway (Fig. 4.8).

Such an integration pathway, however, generates a mismatched heteroduplex that needs to be resolved either by replication or repair. In pneumococci, two types of markers, viz. low efficiency (LE) and high efficiency (HE) have been identified. The basis of this difference lies in the *hex*-encoded mismatch repair system. While in LE type, the repair system recognizes and eliminates the introduced marker, HE markers escape such mismatch correction. Hex mutants that are expected to show HE nature for all the markers have been isolated and lend support to this theory.

Coumermycin, a DNA gyrase inhibitor inhibits the stable donor–recipient complex formation in *B. subtilis*, suggesting that superhelical molecule may be preferred for recombination. Similarly, a drop in temperature from 37 to 17 °C halts integration and transformation remains sensitive to S1 nuclease (single-stranded DNA-specific). In *Haemophilus* competent cells, single-stranded gaps are present in the recipient DNA, which may facilitate donor–recipient complex formation.

In all the three systems, mutants of *rec* type fail to integrate the DNA, suggesting some common steps with recombination. That these mutants are also sensitive to DNA damaging agents provide further support.

4.5 Artificial Transformation

All the systems described above are classified as naturally transformable. There are, however, many bacteria which cannot do so. One classical example that immediately comes to mind is that of *E. coli*, whose genetics developed rapidly during this period, providing a lot of inputs in terms of genetic markers to start extensive genetic analysis, and molecular and functional characterization. However, the deployment of *E. coli* in many genetic studies and more importantly in gene cloning experiments necessitated the development of a transformation system, a system often referred to as artificial transformation. The main difference between the natural and artificial transformation system is based on the induction of competence in the latter as compared to the acquisition of the same in the former. There are different ways in which the competence \ transformation can be effected in these systems, such as (i) treatment with certain chemicals, the commonest being divalent cations, (ii) protoplast transformation using polyethylene glycol (PEG), (iii) electroporation, and (iv) biolistic transformation.

4.5.1 Chemical-Induced Competence

After the initial work of Mandel and Higa (1970), who showed that *E. coli* cells treated with high concentration of $CaCl_2$ at low temperature can be transformed with plasmid and phage DNA, a whole lot of experiments were designed to make transformation a workable technology in those bacteria that are naturally non-transformable.

Different *E. coli* strains respond differently to various competence raising conditions. The most popular, however, is the treatment of cells with multivalent cations such as, Ca at 0 °C, though many other cations like, Mg, Mn, and Ba may actually be better. In many other strains, treatment with organic solvents, sulfhydryl reagents, and hexamine cobalt chloride have given improved transformation frequency. In some others, repeated freezing–thawing may also bring about the desired alterations.

The thick outer membrane in Gram-negative bacteria is regarded as one of the impediments to the access and transport of DNA. It has been found that growth in 20 mM Mg reduces the number of protein–LPS bonds by increasing the number of ionic bonds over covalent bonds. This increases the fluidity of the membrane, and in turn facilitates the transformation process. Similarly, growth in rich medium improves transformability as also the growth at lower temperatures (25–30 °C) as compared to 37 °C. At lower temperatures, cell synthesizes membranes with fewer saturated lipids thus increasing its fluidity. The role of LPS is also borne out from the observation that LPS with shorter O-side chains are more effectively transformed perhaps because it improves the accessibility of the DNA. Standardized protocols and strains of *E. coli* are available to carry out high efficiency transformations; in fact, competent *E. coli* cells are also commercially available.

The phospholipids, a variety of proteins, and LPS provide the cells a net negative charge. DNA with its phosphate backbone is also anionic. Thus, any interaction between them could be considered unfavorable. These impediments are supposed to be overcome by $CaCl_2$ treatment at low temperature. While low temperature (0 °C) may freeze the distribution of phospholipids, LPS, and proteins on the cell surface producing a quasi-crystalline structure, Ca (or other cations) will shield the phosphates on the cell surface as well as approaching DNA. This creates a favorable condition for DNA transfer across the membrane.

The possible route of DNA entry into *E. coli* cells is through channels known as zones of adhesion or Bayer's junction. A typical *E. coli* cell may carry as many as 400 such zones of adhesion. The role of these channels was established when cobalamine which also uses these channels was found to competitively inhibit DNA uptake. Another type of channel implicated in DNA uptake consists of poly (HB):poly P:Ca. In this poly (HB) is envisioned to wrap around DNA (itself a poly P), and is carried in a shield formed by Ca ions.

No sequence-specific DNA association has been detected in *E. coli*. In fact, its non-selectivity toward taken up DNA has been thoroughly exploited in gene cloning experiments, making *E coli* if not the most favorable, at least often the first host to receive the foreign DNA. The double-stranded DNA is the preferred form, and the transformation efficiency decreases linearly with the increasing size of the plasmid, with very large plasmids finding it difficult to be taken up intact. Plasmids (ccc DNA) fare better as transforming DNA than the linear forms. The average size of the taken up DNA is usually more than the cell size or channel size. Thus DNA must be compacted during entry. Competent cells are capable of taking multiple DNA molecules unless the number of plasmids far exceeds the cell number. A short heat pulse (0 °–42 °–0 °C) appears to complete the DNA uptake process perhaps by transiently melting the crystalline membrane. A brief interlude of good growth conditions allows the introduced DNA to establish itself in the new cellular environment. Following the growth, cells are plated on selective medium to score for the transformants.

Two general classes of genetic restriction functions may serve to monitor any DNA that enters the cell: (i) the classical K-type restriction-modification (hsd RMS) that restricts unmethylated DNA and has perhaps evolved to control bacteriophage infections, and (ii) three methylation-dependent restriction systems (MDRS) that may represent a component of natural mechanism of DNA transformation designed to allow limited diversification.

Many Gram-negative bacteria have now been successfully transformed making transformation a universal experimental tool to transfer desired genes into bacteria.

4.5.2 Polyethylene Glycol (PEG)-Mediated Transformation

Many Gram-positive bacteria that do not acquire competence naturally as also those which do, can be transformed by protoplast transformation using polyethylene glycol 6000 (PEG). In this method, peptidoglycan cell wall is removed and the resulting protoplasts are maintained in an osmotically stabilized condition. The protoplasts are then subjected to donor DNA, plasmid or phage in presence of PEG 6000. Although the role of PEG is not very clear, it is known to fuse the cell membrane, perhaps trapping the DNA in close contact and facilitating its uptake. After DNA uptake, the cell wall needs to be regenerated before the transformants can be selected. Although of a wider applicability, isolation, maintenance, and regeneration of protoplast requires specific care and needs to be standardized for each system. Plasmid and phage DNA fare better than the chromosomal fragments.

Besides the methods described above, two novel techniques of wider applicability have been designed to introduce nucleic acids into a cell with improved efficiency. These are: electroporation, and biolistic, high-velocity projectile-mediated DNA transfer. Both have been successfully applied to eukaryotic cells and the latter has in fact been established for eukaryotes.

4.5.3 Electroporation

Electroporation or electrotransformation is one of the most efficient methods of introducing DNA in a cell. Although designed for eukaryotes to begin with, it works with equal success in prokaryotic cells. Essentially, it consists of passing a high voltage electric discharge in brief pulses through a suspension of cells and DNA. Electric pulse is supposed to transiently disorganize the membrane and thus permeabilise the cells toward the entry of a variety of macromolecules including DNA. The strength of the electric current and length of the pulse needs to be optimized for effective permeability and productive association with DNA. For each bacterial strain, optimal electroporation range is first identified and for many model systems standardized protocols are now available. Growth of the cells in a medium lacking Mg is shown to improve the efficiency. Also, the cells must be thoroughly washed to remove all traces of salts. In generalized protocols of electroporation, high cell density (5×10^{10}–1×10^{11}) suspension is used as high level of killing is expected during the process. These cells are maintained at low temperature so as to overcome the adverse effects of any temperature changes that occur during the passage of electric current. Transferring the cells to the growth medium maintained at the optimum temperature also improves the transformation frequency by ten-fold. Under optimum conditions, the transformation efficiency of as high as 10^9–10^{10} transformants per microgram of DNA can be obtained.

The mechanistic details of the process do not vary much from that in the chemical-induced competence.

4.5.4 Biolistic Transformation

This method involves the acceleration of DNA precipitated on to spherical gold particles into eukaryotic tissue with a specialized device. Basically, it consists of bombardment of gold–DNA mixture from a distance onto the target tissue that is densely packed on filters. The cells are then incubated and regenerated for selection. However, it is still not a very popular method, more so with prokaryotes, where other methods are more practical.

4.6 Transformation in Genetic Analysis

Besides being a mechanism of natural gene exchange in many bacteria, and its development as a generalized method of gene transfer as also its applicability in recombinant DNA technology as a step for introducing DNA, transformation has served the purpose of doing the linkage analysis in bacteria. This is based on the contention that as a large piece of DNA can be inserted, that may carry more than one gene. Moreover, the closer the genes (linkage) more would be the chances of them being taken together, leading to what is known as cotransformation.

In this process, a recipient population, mutant for more than one marker can be transformed with the donor DNA, carrying the complementary wild-type alleles. The transformants for two single, as well as both the markers or cotransformants can be scored (Fig. 4.9). The frequency of cotransformation is then compared with the single transformation frequency for the two markers. Thus, if the two genes are unlinked and therefore, carried on two different DNA fragments, the probability of getting cotransformation would be equal to the product of two single transformation frequencies. However, if the genes are linked and thus are located on the same DNA, the cotransformation frequency will be significantly higher than the product of the two, and in fact, may be very similar to the single transformation frequency (Fig. 4.9). One must also remember that this relationship can be applied only if the population consists of 100 % competent cells, if not, the frequencies will have to be appropriately corrected. For example, if competent a b cells are transformed with a^+b^+ donor DNA, three different types of transformants viz., a^+b, ab^+, and a^+b^+ can arise. If the frequency of a^+b transformant is 1/100 and that of ab^+ is also 1/100, the frequency of a^+b^+ (co-transformant) will be expected at 1/1,000 cells if the two genes are unlinked but significantly higher than 1/1,000 if the genes are linked. You may recall that this type of calculation is based on the assumption that 100 % cells are competent. If not, as in *B. subtilis*, where only 20 % cells are competent, the same frequency of a^+b^+ unlinked transformant class will be 1\400 instead of 1/1,000 cells.

There is another way to differentiate between the linked and unlinked transformants. In this, the competent ab bacteria are exposed to decreasing concentrations of a^+b^+ donor DNA. You may recall that this will lead to a corresponding decrease in the number of transformants. If the two genes are linked, the drop in a^+b^+ cotransformation frequency will be the same as the fall in either a^+b or ab^+ single

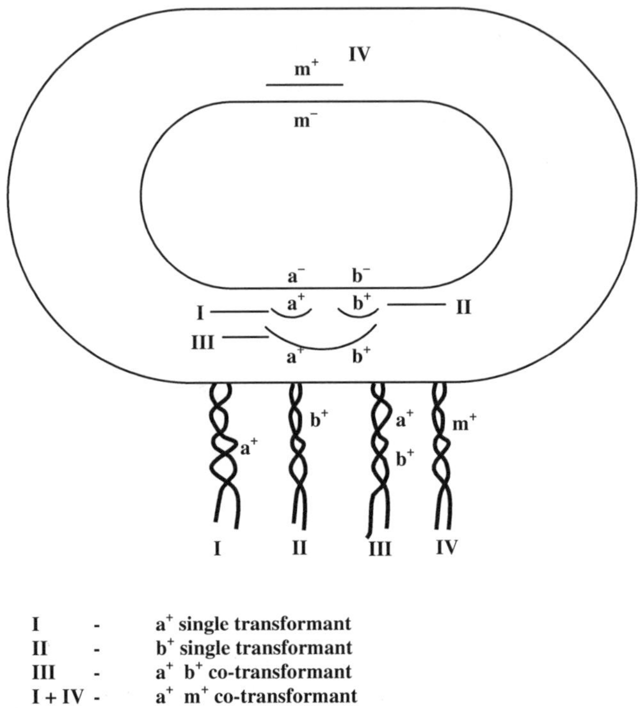

Fig. 4.9 The formation of single and cotransformants. While the single transformants are formed due to uptake and assimilation of complementary wild-type DNA (I, II, and IV), a cotransformant could be due to two genes being linked (carried on the same piece of DNA-III) or simultaneous uptake of two different unlinked pieces of DNA (I + IV). If: cotransformation frequency $= (a^+) \times (m)^+$—genes unlinked (a^+ m^+). Cotransformation frequency $> (a^+) \times (b^+)$—genes linked (a^+ b^+)

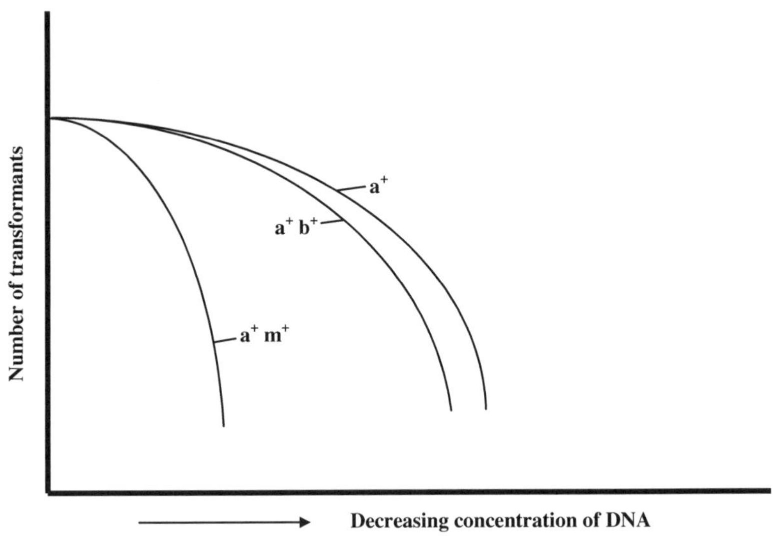

Fig. 4.10 Differences in the cotransformation frequency as monitored by transforming recipient population with decreasing concentrations of donor DNA. *Note* if the genes are linked (a^+ b^+) the fall in the number of transformants follows a pattern similar to that of single transformant, e.g., a^+. If unlinked, however (a^+ m^+) the fall would be sharper

transformants. However, in case the two genes are carried on different DNA fragments (unlinked), the slope of the curve will be different and much sharper (Fig. 4.10).

Although the application of transformation in linkage analysis is being practiced on a regular basis, in this era of genome sequencing it has lost much of its relevance as linkage of markers can also be deduced by physical maps.

The other application of transformation is in strain construction. Transformation, unlike conjugation, is a method of choice if a small defined modification is required in a chosen genome. And as a crucial step in the recombinant DNA experiments, this method has a wider application of introducing desired foreign DNA into a cell.

Further Readings

Avery OT, Macleod CM, McCarty M (1944) Studies on the chemical nature of the substance inducing transformation of pneumococcal types. I. Induction of transformation by a deoxyribonucleic acid fraction isolated from pneumococcus type III. J Exp Med 79:137–158

Bergmans HE, van Die IM, Hoekstra WP (1981) Transformation in *Escherichia coli*: stages in the process. J Bacteriol 146(2):564–570

Campbell EA, Choi SY, Masure HR (1998) A competence regulon in *Streptococcus pneumoniae* revealed by genomic analysis. Mol Microbiol 27:929–939

Chen I, Dubnau D (2004) DNA uptake during bacterial transformation. Nat Rev Micrbiol 2:241–249

Etchuuya R, Ito M, Kitano S, Shigi F, Sobue R, Maeda S (2011) A novel type of natural transformation involving cell-derived DNA and a putative promoting pheromone. PLoS ONE 6(1):e16355

Kaufenstein M, van der Laan M, Graumann PL (2011) The three layered DNA uptake machinery at the cell pole in competent *Bacillus subtilis* cells is a stable complex. J Bacteriol 193(7):1633–1642

Lacks SA (2003) Rambling and scrambling in bacterial transformation—a historical and personal memoir. J Bacteriol 185(1):1–6

Lorenz MG, Wackemagel W (1994) Bacterial gene transfer by natural genetic transformation in the environment. Microbiol Rev 58(3):563–602

Mercenier A, Chassy BM (1988) Strategies for the development of bacterial transformation systems. Biochimi 70(4):503–517

Petersen FC, Tao L, Scheie AA (2005) DNA binding-uptake system: a link between cell-to-cell communication and biofilm formation. J Bacteriol 187(13):4392–4400

Radfield RJ, Schrag MR, Dean AM (1997) The evolution of bacterial transformation: sex with poor relations. Genetics 146:27–38

Steinmoen H, Knusten E, Havarstein LS (2002) Induction of natural competence in *Streptococcus pneumoniae* triggers lysis and DNA release from a subfraction of cell population. Proc Natl Acad Sci U S A 99:7681–7686

Wirth R, Friesenegger A, Fiedler S (1989) Transformation of various species of gram-negative bacteria belonging to 11 different genera by electroporation. Mol Gen Genet 216:175–177

5 Transduction

5.1 Discovery

Transduction was first discovered in 1951, in *Salmonella typhimurium* (=*S. enterica* serovar Typhimurium) by Zinder and Lederberg. Their experiment was directed toward establishing an *E. coli*-like conjugation mechanism in this bacterium. The protocol thus followed was to mix two multiple auxotrophic strains and look for prototrophic recombinants. Although such recombinants were produced, neither a cell-to-cell contact (conjugation) was required nor was the process affected by the treatment with DNase (transformation). They, therefore, called this process transduction, and referred to the product as a transductant. The basis of this process was assigned to a filterable agent (FA), which was later demonstrated to be identical to the bacteriophage P22. The comparison between the two was based on a number of characteristics, such as: the dimension and the mass of the two were similar; both were resistant to DNase; both could be inactivated by heat and antisera raised against them; and the host range of both was the same, such that phage-resistant mutants of the host failed to interact with FA. Then in 1955, Lennox described a similar transduction in *E. coli* by the phage P1. In both the cases, transduction for a large number of donor markers could be demonstrated. Another coli phage lambda (λ) was also found to indulge in such a process but contrary to phage P22 and phage P1, it had a very limited capability. These three examples clearly established that a bacteriophage can act as a carrier of bacterial DNA from one cell to another. Since these initial discoveries, transduction has turned out to be a process widely occurring in the bacterial kingdom.

The transduction carried out by phage P22 and phage P1 is termed generalized transduction to differentiate it from the phage λ-mediated process that came to be called as specialized or restricted transduction. While in the former, practically any gene can be transferred from one bacterial cell to another by a phage, in the latter, such genes are few and limited. However, as we shall go into the details of the two processes, it may become clear that the only commonality between these processes is their mediation by a phage. In most other details, such as, in the generation of transducing particles, their genetic composition, and the formation of transductants, the two differ widely.

5.2 Vegetative Growth of Phage

It is during the multiplication (vegetative growth) of a phage inside the bacterial cell that a phage may deviate from its course in producing the phage progeny (infectious phage) only, to a mixture of infectious and transducing phage (the latter carrying the host DNA). It is thus imperative to understand the steps involved in a phage life cycle, so as to know when and how such a deviation may occur.

S. Srivastava, *Genetics of Bacteria*, DOI: 10.1007/978-81-322-1090-0_5, © Springer India 2013

Basically the phage multiplication or vegetative growth comprises the following steps:

To begin with, the phage makes a productive association with a specific host by recognizing certain receptor sites, which is followed by a proper attachment or adsorption. Thereafter, the phage through an intricate mechanism creates a clear passage between its head, where its genome is lodged, and the inside of the host cell (cytoplasm). Now the phage nucleic acid is injected\transferred into the cell cytoplasm. Subsequent to this, the phage nucleic acid takes over the metabolic control of the host cell, diverting the host machinery and the resources to synthesize several copies of its own genome, as well as directing the synthesis of all the required capsid proteins. Once the capsid proteins are synthesized, they are assembled in a characteristic fashion to produce different phage structures, such as head, tail, tail fibers, base plate, and such other structures depending upon the phage concerned. Phage head is now filled (packaged) with a copy of its genome, and then other structures join the head in a process referred to as phage morphogenesis to produce what is known as a phage particle or a virion. Once the designated number of phage particles or progeny (burst size) becomes available, the phage lyses the host cell and releases the phage population (Fig. 5.1). These phages have no extracellular life and thus must find a new host cell to repeat the same steps as outlined above. After two to three such rounds of infection-lysis, a visible sign of host cell killing is produced in the background of a growing lawn of bacteria. This sign is referred to as a plaque, the shape and size of which is a phage determined genetic character. Each plaque, in fact, reflects the initial contact between the phage and a host cell, and plaque counting (plaque forming units or PFU/ml) is used to titer the phage population. Such phages are capable of multiplying, once they find another host cell, and thus are also referred to as infectious phage.

This phage growth cycle often referred to as the vegetative growth is the time spent by the phage inside the host cell to complete the infection process (to multiply), and will also have to be employed if a transducing version is to be generated. We can now discuss the two pathways of transduction in detail.

5.3 Generalized Transduction

As described earlier, one type of transduction mechanism is referred to as a generalized one. Although many phages may support this mechanism, the two model systems identified are of phage P22 and P1. In order to understand the role played by these phages, it is important to know their structure as it helps in unraveling how a phage picks up a piece of bacterial DNA while it multiplies, so as to serve the purpose of transduction.

P1 is a typical phage consisting of an icosahedral head, an intricate tail with contractile sheath, base plate, and tail fibers. P22 in comparison is much smaller and simpler in organization. The tail is so reduced that the head appears to rest directly on the base plate. The smaller size of phage P22 is also reflected in a smaller genome ($\sim$44 kb) in comparison to P1 which carries a genome of $\sim$100 kb. Thus, P1 genome is $\sim$2 % of the host *E. coli*, and phage P22 is $\sim$1 % of the host *Salmonella* genome. The other characteristic of the two genomes is their terminally repeated/redundant and circularly/cyclically permuted ends. This is extensive in phage P1 ($\sim$12 %) compared to phage P22, which carries only $\sim$2 % cyclic permutations. In other words, the DNA in individual particles not only differ at the ends, but also the single-stranded DNA from individual particles can reanneal, with terminal redundant permuted ends appearing as unannealed tails or bushes (Fig. 5.2).

5.3.1 Formation of Transducing Phage

In 1965, Ikeda and Tomizawa established that transducing phage particles are produced by the same set of bacterial cells in a population that principally produces the infectious phages. They

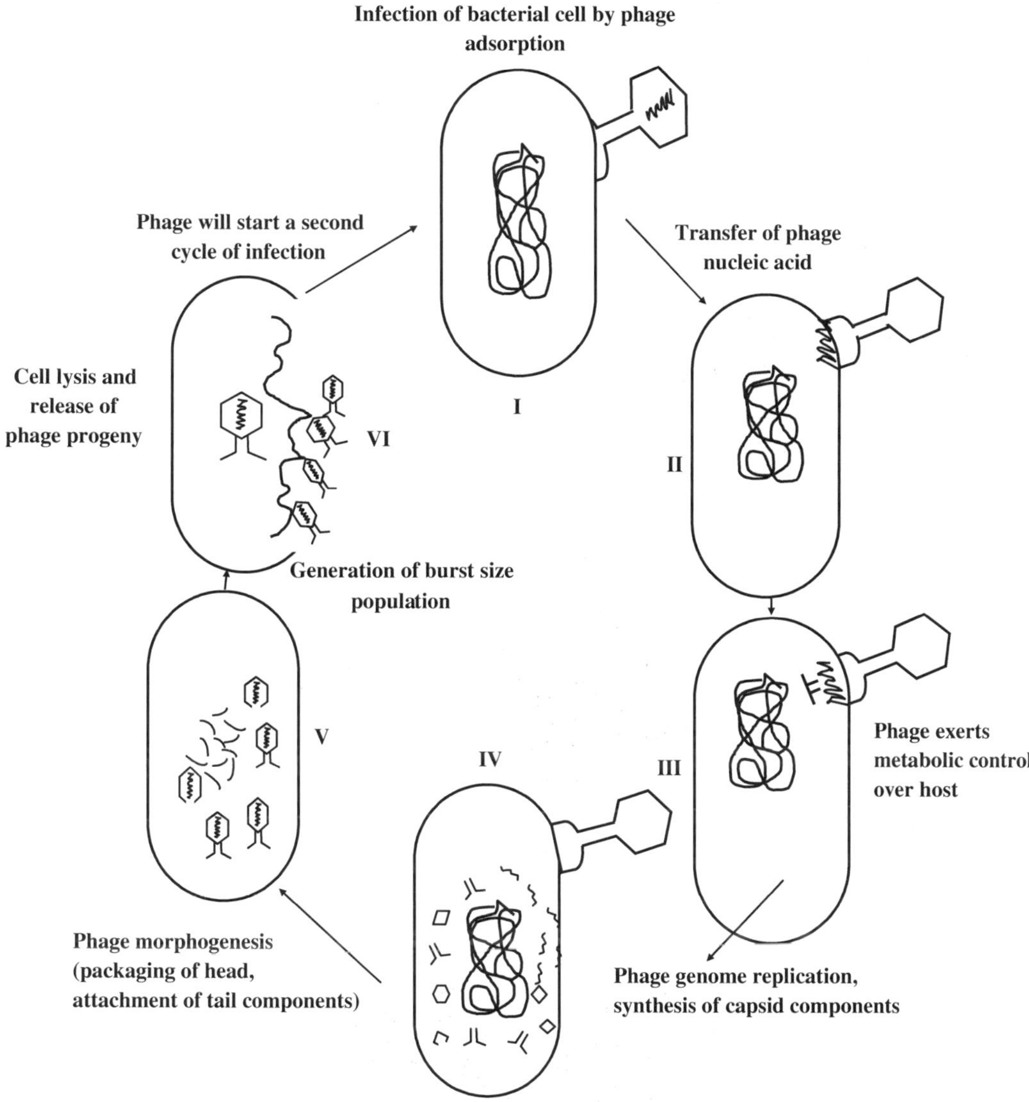

Fig. 5.1 Different steps in phage infection cycle leading to phage multiplication (Vegetative growth)

investigated the formation of P1 transducing phage by labeling of *E. coli* DNA with 5-bromouracil (5-BU), a heavy analogue of thymine, and tracing its fate in three different sets (Fig. 5.3). In one control set, *E. coli* DNA was labeled before as well as during infection, and in another control, it was not labeled at all and was thus normal. In the third set, host DNA was labeled with 5-BU before phage infection but the cells were shifted to normal thymine medium during infection. The basis of this experiment was to differentiate DNA that is synthesized before infection (of host) and the one that is made afterwards (of phage). The phage particles were collected from each set, differentiated into two types (infectious and transducing) and analyzed through density gradient centrifugation. On the basis of DNA density labeling pattern, they observed that while in the first two cases, the DNA of infectious and transducing phage conformed to either the heavy or normal density, the third case showed a different profile. In the latter class, while the DNA of the infectious particles was normal that of transducing phage

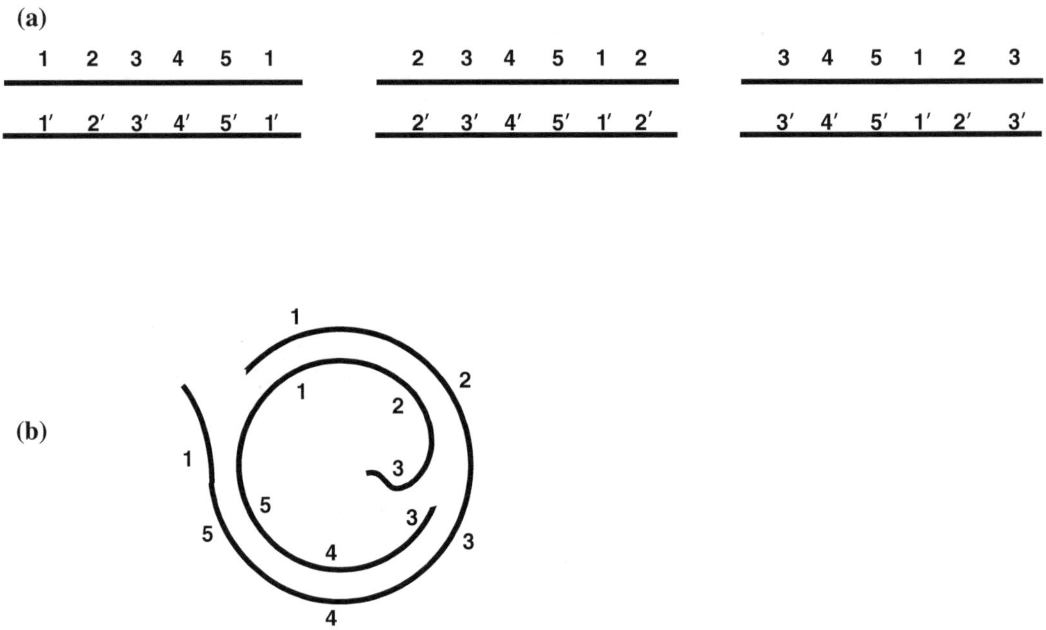

Fig. 5.2 **a** Terminally redundant and circularly permuted ends of phage genome. Note: The individual particles differ at their ends. **b** When single-stranded DNA from individual particles are annealed, their permuted ends may appear as unpaired tails

was heavy. They could draw two major inferences from these results. First, the two categories of phages do not carry a mixture of DNA (host and phage), thus producing these two distinct types of particles separately. Second, the transducing DNA is derived from the existing host DNA and no prior replication is required (Fig. 5.3). In the initial experiments, the phage P1 strain used was a low transducing type, producing ~0.3 % of the transducing progeny in a total phage population. In some of the later work, this was replaced by strains, where this value could be raised to ~2 % of the population. P22 essentially resembles phage P1 in that the transducing phage carries only bacterial and the infectious particles only phage DNA.

It thus appears that transducing particles are produced by mistaken packaging of the host DNA, such that the phage machinery may seize upon the bacterial DNA as its own and package the same. This looks plausible as the host cell cytoplasm where phage morphogenesis is going on is also interspersed with bacterial DNA.

5.3.2 Packaging of Phage DNA

The two model systems of phage P1 and P22, have the characteristic terminally redundant ends with circular permutation. On the basis of this similarity, it was envisaged that P1 and P22 DNA are packaged from a concatamer of several genome equivalents, and that the packaged DNA, though of a fixed length is longer than the unit genome. The consequence of such a mechanism is that the phage has a sequence repeated at each end. Moreover, since subsequent molecules are cut out from the same concatemer, these ends are all different, showing circular permutation (Fig. 5.4). The extent of circular permutation depends upon the number of genome lengths derived from a single concatemer. If this is limited as in case of phage P22, it may suggest that there is a preferred site where packaging begins and that it goes on for a few cycles.

That these two phage systems do possess this type of packaging mechanism was confirmed

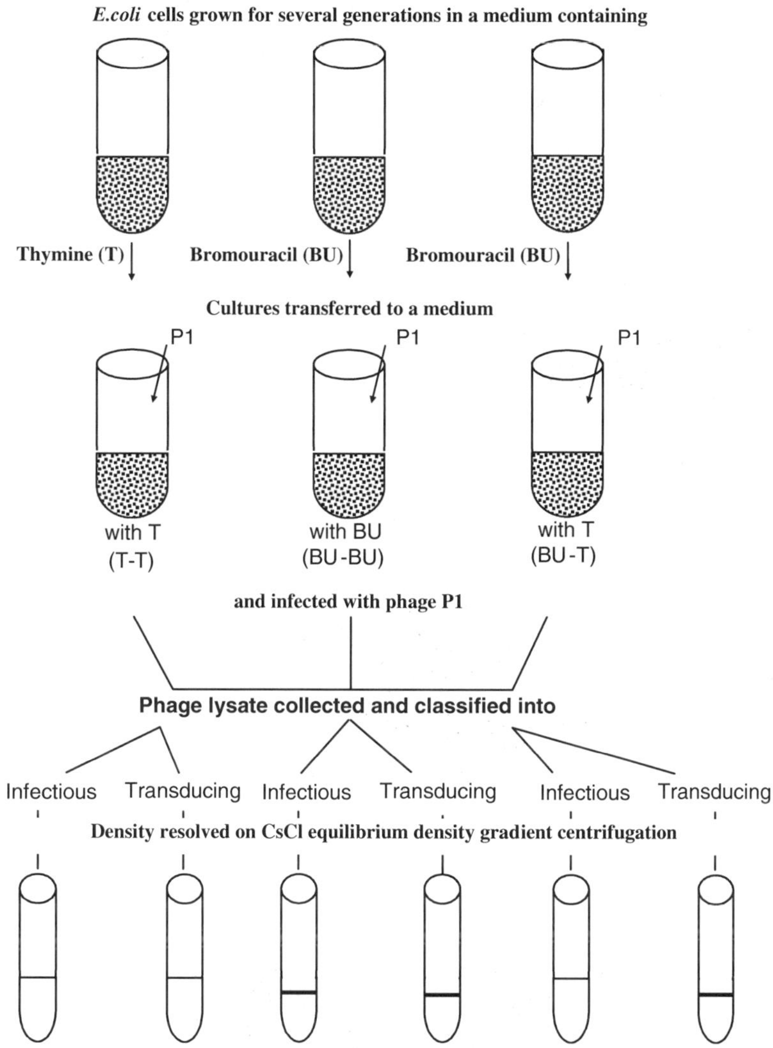

Fig. 5.3 Ikeda and Tomizawa's experiment to show the type of DNA carried by infectious and transducing particles of phage P1 growing on *E. coli*. Note that Bu-T transducing phage has a density heavier than infectious particles

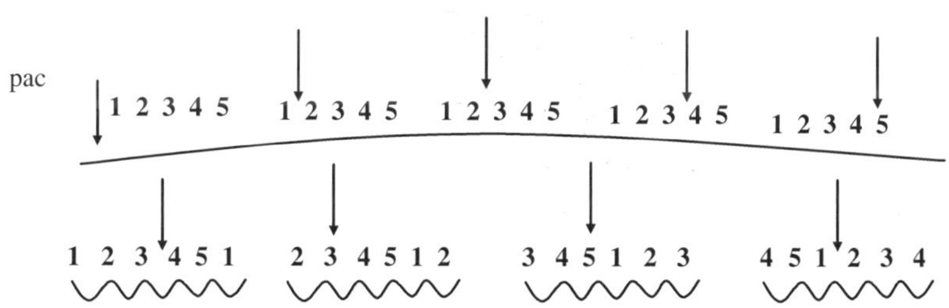

Fig. 5.4 Generation of terminally repeated and circularly permuted ends of phage (P22 and P1) genome. The *upper* panel shows the creation of concatemers of genome length by replication. Phage genome is packaged by head-filling mechanism that requires a length longer than the unit genome. On the *upper* panel ↓ suggests the cutting sites for packaging

when their variants were produced, carrying either an insertion of an extra length of DNA in the genome or had a part of the genome length deleted. This led to either increased terminal repetitions or the same is completely abolished. Also, since the ends of the genome are permuted, the length of the packaged DNA cannot be determined by the sequence at the ends. The generally accepted mechanism is of "headful" packaging, in which DNA is packaged from a concatemer and in which genome length capable of head filling is longer than the unit genome. In phage P22, packaging is initiated at a designated site, called *pac* and then proceeds by headfuls. In phage P1 a similar mechanism operates.

DNA packaging in phage P22 is initiated in a region of ~120 bp that contains *end* sites (6 in number) and a *pac* site that is used as a recognition site. Six *end* sites are separated by 20–40 bp, and the *pac* consisting of 17 bp lies in the center of the *end* region. Encapsidation (packaging) also requires five phage-encoded proteins. Of these, three constitute a part of the prohead structural components and the remaining two, gp2 and gp3, are important and are transiently associated with the prohead. Both gp2 and gp3 recognize the *pac* and cut at the *end* site. In their absence, both proheads and DNA concatemer accumulate. DNA possibly still associated with gp2 and/or gp3 enters the prohead. As the prohead matures, a nuclease (perhaps one of the portal proteins) cuts at the other end of the DNA. Three more virion proteins are required to stabilize and lock the DNA inside. Then gp2 and gp3 leave the virion and the tail attaches.

In phage P1, products of two co-transcribed and co-translated, *pacA* and *pacB*, are required for packaging. The process is initiated in one of the *pac* genes. The functional cleavage site is contained within a 162 bp segment and cleavage occurs centrally within this sequence. The *pac* (cleavage recognition region) contains seven repeats of TGATCA, three to the right and four to the left of the cut site. The GATC sequence in the target is methylated either by the phage or host methylase. This was confirmed by showing a highly reduced growth of a Dam phage in a *dam* host. The DNA of the transducing phage P1 is associated with a protease sensitive component.

Early experiments had clearly shown that both the infectious (PFU) and transducing particles are produced by the same host cell population. Thus, the only probable mechanism appeared to be the mistaken encapsidation of the host DNA instead of that of the phage in producing the transducing particles (Fig. 5.5). The early thought of the host DNA fragmentation to the phage genome-sized fragments could not be confirmed suggesting that the entire genome serves as the substrate for encapsidation. Wild-type phage P22 packaging machinery shows over 40 times more preference for phage DNA than the bacterial DNA. Wild-type phage DNA, therefore, has better initiation signals than the host. Also, P22 transduction frequency for different bacterial markers varies as much as 1000-fold, leading to the suggestion that every marker is not available for encapsidation at the same probability, once again indicating that there are preferential sites for initiation of packaging. It has been shown that *Salmonella* chromosome has several sites dispersed all over its genome that bear sufficient resemblance to P22 *pac* sites (Fig. 5.6). These *pac*-like sites, thus can be recognized by packaging enzyme and employed for processive packaging series. If such sites are limited, any deletion or insertion will change the register of packaging series and the frequency of co-transduction (Fig. 5.7).

P22 HT mutants (high transduction) show 100- to 1000-fold enhanced transduction that results from enhanced encapsidation of host DNA (as much as 50 % of host DNA utilization). These mutations map to gene3 and mutant phenotype is dominant or co-dominant. Mutations, however, might also alter or reduce the packaging specificity or generally increase the level of cutting. When a phage *pac* site is introduced into the host chromosome, the transduction frequency of the downstream markers may improve up to 10-fold. With phage P1, transduction frequency for different *E. coli* genes showed comparatively little variation, suggesting that host DNA may be packaged by

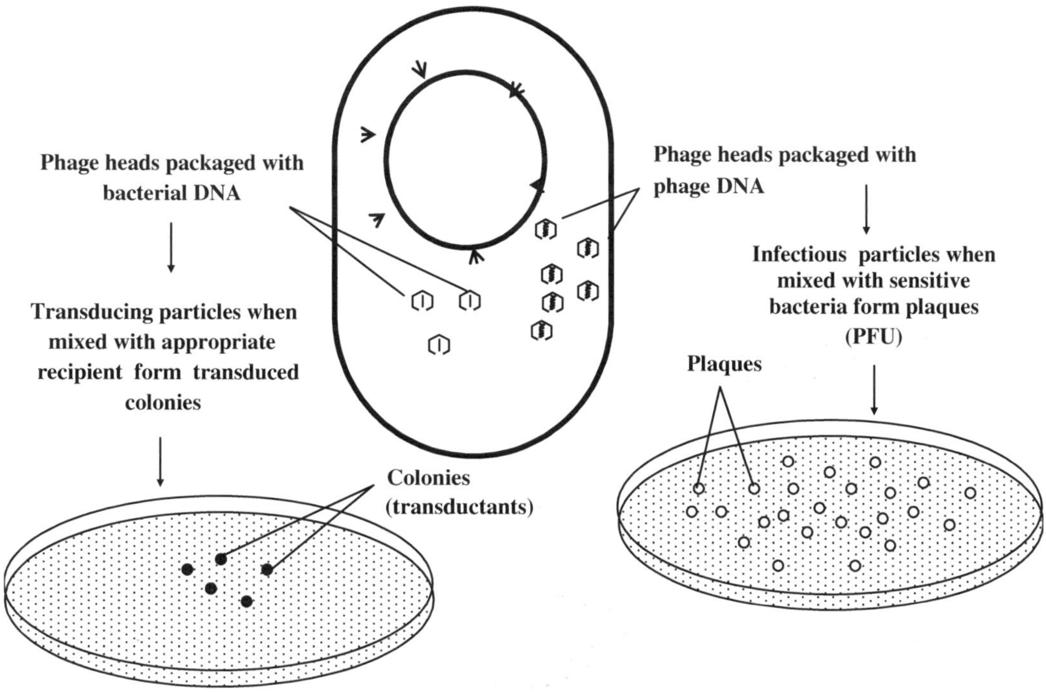

Fig. 5.5 Generation of infectious (PFU) and transducing particles. While normal packaging will result in infectious particles or PFU, the mistaken packaging will generate transducing particles carrying literally any part of the bacterial chromosome. Both events may take place in the same bacterial population but not in the same cell

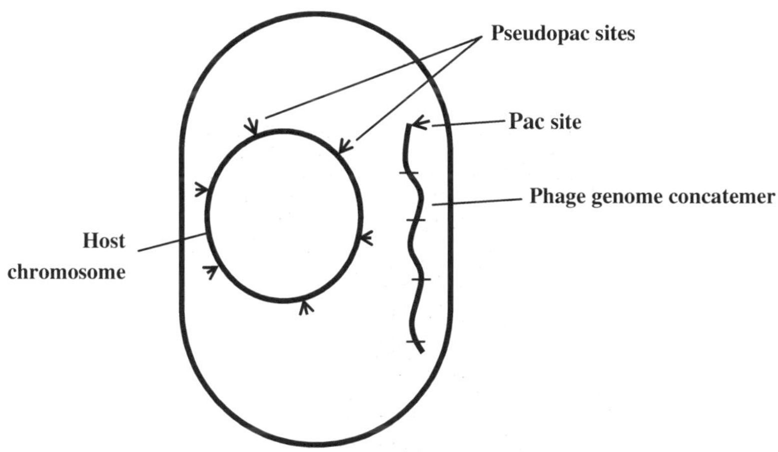

Fig. 5.6 Formation of a generalized transducing phage. Phage P22 and P1 package their genomes by recognizing a specific *pac* site. When mistaken packaging takes place, phage P22 packaging machinery recognizes pseudopac (*pac*-like) sites on *Salmonella* chromosome and uses it as a packageable substrate. For phage P1, it starts non-specifically as *E. coli* chromosome does not carry such *pac*-like sites

phage P1 in a sequence non-specific manner either from DNA ends or through a non-specific cutting by the packaging enzyme. That the *E. coli* chromosome does not carry a *pac*-like site was proved by the lack of hybridization between a *pac* DNA probe and the host chromosome. However, when a *pac* site is inserted into *E. coli* chromosome by transposition,

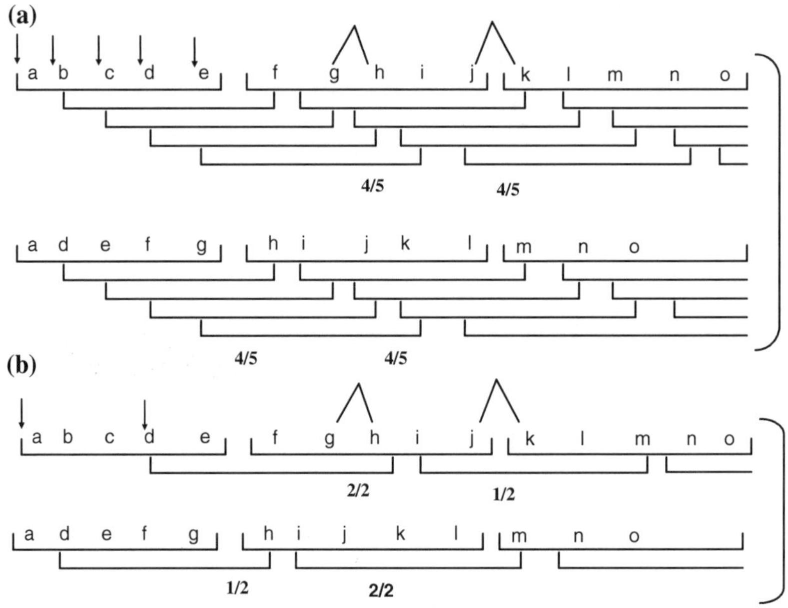

Fig. 5.7 Sequential packaging from a DNA substrate with either (**a**) non-specific or (**b**) specific cutting site (↓), and the consequent effect of a deletion in the sequence. While the co-transduction frequency (shown here as for *g* and *h*, and *j* and *k*) in the former will not change, in the latter it will show a distinctive difference

unidirectional packaging of bacterial DNA is initiated from this site for at least five headfuls at 100 % efficiency, and thereafter for an additional five headfuls with a distance-dependent decrease to about 5 % efficiency. Sequential packaging even if it initiates in a non-specific manner, has been demonstrated by processive packaging of three different phages (λ, 21, and 186) from a single burst. P1 HT mutants produce 3–5 times as many transducing particles as the wild type. These mutations cannot be complemented by co-infection with a non-mutant phage suggesting that they do not lack the required enzyme. Enhanced chromosomal breakdown appears to be the basis that may create extra ends for packaging.

5.3.3 Formation of Transductant

Once the transducing phage, carrying a piece of host DNA instead of its own, is produced and released, the phage must find an appropriate new host cell to deliver this material. The host cell is now referred to as recipient that will utilize this information to produce the transductant. In the most ideal situation, the recipient must carry different genetic information so that the final product or the transductant can be easily selected or distinguished. In fact, once the phage infects the recipient and transfers the donor DNA into cell cytoplasm, any of the three pathways may decide its fate: (1) recombination with the chromosome to give complete transductant; (2) donor DNA destruction and reutilization of the components; and (3) neither integration nor disintegration of donor DNA, a situation referred to as abortive transduction.

Of the three, the formation of abortive transductants is the most frequent. This was first demonstrated by certain genetic experiments. For example, when transductants were selected for normal colony growth (seen as producing normal-sized colonies), very small-sized colonies (like that of the recipient used in this experiment) among the normal-sized colonies were also detected. Analysis of individual cells in the transduced colonies revealed that each contained one cell that could give rise to another small colony. The explanation for this behavior is that when transduced DNA fails to recombine with the homologous recipient DNA, it aborts. Such a DNA remains in the cell cytoplasm, may be expressed, and even complement a mutation in the recipient. When such a transduced cell divides, free donor DNA fragment fails to replicate and thus goes to only one of the two daughter cells. The protein product, however, is passed on to both the cells allowing the growth of even that cell which does not inherit the donor DNA. In such a cell, synthesis of the required protein will not take place so that the said protein will eventually dilute out, and as a result the growth will also stop. This leads to the small-sized colony as observed. The cell that retains the original donor DNA can reinitiate this process and give rise to another such colony (Fig. 5.8).

In order to determine the basis of abortive transduction, donor DNA was re-extracted and its fate was studied. For this, donor DNA was density labeled with 5-BU and it was found that the DNA remained in the heavy fraction, apparently unassociated with the recipient genome. This suggests that donor DNA under abortive conditions neither degrades nor recombines but remains free, a state that may last as long as 5 h in *E. coli.* Molecular analysis of DNA based on its density and gel mobility, showed it to be a circular molecule. Circularization was not easily explainable as the ends did not carry any terminal redundancy or any other appropriate feature to effect circularization. Treatment with protease or other protein-dissociating agents was found to convert this circular structure into a linear one indicating that a protein is involved in converting the donor DNA fragment into a circular form.

A lot of work has been done on the nature and origin of this protein. Ikeda and Tomizawa, while working with phage P1, described a protein that remains associated with the transducing DNA. They proposed that this protein originates in the donor cell, packaged along with the DNA and could carry out this protective circularization step. Yarmolinsky and Sternberg suggested that protecting protein is in fact a packaging endonuclease that remains associated with DNA. Two internal head proteins of phage P1, DarA and DarB, have also been implicated in the protective function. In phage P22, HT mutants having high frequency of complete transduction have been isolated. One class of such mutants fails to yield abortive transductants. The mutation has been mapped to gene 16, whose product is packaged along with DNA. The protein seems to protect the ends by bringing them together.

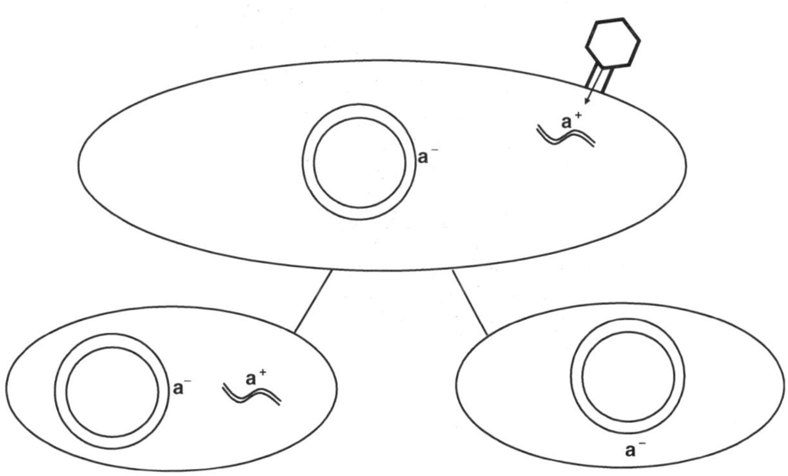

Fig. 5.8 Abortive transduction—when the donor DNA does not integrate into recipient genome, the marker carried on the same cannot be stably inherited, though the phenotype may be expressed. At each cell division, one cell which does not receive the donor marker will show non-transduced (like recipient) phenotype

Abortive transductants, however, rarely get converted to complete transductants. In other words, such circular molecules are not destined to recombine.

Integration or recombination that gives rise to complete transdunctants takes place within the first 1 h of the donor DNA transfer and requires *rec* function (see Fig. 5.12). On the other hand, in ~10 % of the cases the DNA does not recombine but gets degraded. This happens in both rec^+ and rec^- cells. Irradiation of transducing phage before infection increases the proportion of the complete transductants. This stimulation was more apparent for those markers that are otherwise poorly transduced. Thus, damaging the donor DNA may provide a better recombinational substrate, corroborating the fact that differences in P1 transduction frequency may in fact depend upon differential recombination efficiency. Moreover, switchover from RecBC to RecF pathway also appears to support transduction.

5.4 Specialized or Restricted Transduction

Joshua Lederberg and his group, while working with conjugation in *E. coli*, were able to show that in fact another type of transduction mechanism may also exist. One that we have discussed above is the generalized transduction, and the other, specialized or restricted transduction was discovered by them in 1956. Experiments were conducted to test whether phage lambda (λ) found associated with some strains of *E. coli*, is able to transfer the host genes from one cell (donor) to another (recipient), very similar to what was demonstrated with phage P1 or P22 in *Salmonella*. Thus, a wild-type prototrophic *E. coli* lysogenic to phage λ was induced by UV exposure to produce the phage lysate. A variety of non-lysogenic mutant *E. coli* strains were infected with this phage lysate and plated to select for transductants, in which the wild-type allele might have been transferred from the donor to the recipient through the phage. The result of this survey was mostly negative except when the recipient was gal^-. About 10^{-6} of the λ-infected cells did acquire a gal^+ gene from the wild-type donor. This was later extended to show that the *bio* gene could also be transmitted by phage lambda similarly. However, unlike phage P1 and P22, ability of phage λ to carry bacterial genes did appear limited, leading to the mechanism being referred to as restricted or specialized transduction. In order to understand the mechanism of restricted transduction, we must first discuss, as we did under generalized transduction, the way the transducing phage may be generated and take note of differences (or similarities) with the generalized transduction pathway.

5.4.1 Generation of Specialized Transducing Phage

Like phages P1 and P22, phage lambda (λ) is also a temperate phage. However, unlike the former two members, ~50 kb (48,514 bp) genome of phage λ is not flanked by redundant ends, instead both its 5′ ends are extended by 12 nt long single-stranded overhangs. The base sequence of these single-stranded terminal ends are complementary, thus allowing the circularization of the genome once it enters the host cell cytoplasm, followed by the sealing of the nicks by DNA ligase of *E. coli*. These ends are therefore known as cohesive ends or *cos* sites.

Phage λ in the lysogenic mode of life cycle integrates its genome into that of host *E. coli* at a designated site labeled as *att* λ, located between *gal* and *bio* genes. In such a state, referred to as prophage, λ could stay quiescent for a long time until it gets induced, spontaneously or experimentally. Upon induction, the phage genome comes out cleanly from its integrated state, a process known as excision, and initiates the lytic growth that ends in the lysis of the host cell and release of the progeny phage. The integration and excision are mediated by site-specific recombinases (coded by *int* and *xis* genes, respectively). In rare cases (1 in 10^6–10^7 cells), however, this excision may be aberrant, so that instead of the two ends of the integrated λ genome, one cut is made outside, i.e., into *E. coli* genome and another in that of the phage. On the basis of its integration property, we have already seen that host *gal* genes lie on the left and *bio* genes on the right flank. This would mean that depending upon the sites of aberrant excision, *gal* or *bio* genes could be included in the excised genome. In both the events, i.e., phage carrying *gal* or *bio* genes, this will happen with a concomitant loss of a part of the phage genome, as the packaged length of phage genome is fixed (Fig. 5.9). In λ *gal*, the loss will be from the right side of the genome containing some tail and head genes. Hence, λ *gal* phages are defective and are labeled as λ d*gal*. In λ *bio*, the part lost (from the left side) will cover *int*, *xis*, *red*, etc., and thus these particles are apparently normal in their lytic growth and are often referred to as λ p*bio*. With highly truncated genome, rare λ *gal bio* particles can also be generated. DNA packaging like other phages puts a limit on a required length, but particles having 79–106 % of the length are viable.

As imprecise excision can possibly generate only λ *gal* or λ *bio*, this usually restricts the ability of phage λ to pick-up any other gene of the host. Initially producing, the defective λ d*gal* is not a problem as the prophage will provide all the missing essential functions. But perpetuation of such a phage will require a helper phage, either λ^+ or a mutant that can provide the functions λ d*gal* has lost out, at the time of its formation. The size of the bacterial DNA substituting the phage genome in a transducing particle can be determined by conducting crosses between λ *gal* and several normal λ phages, but each carrying different mutations at known locations. If the two do not recombine to give a wild-type λ, the mutant lies in the same region that has been substituted by the inclusion of the bacterial genome. Alternatively, one could also use heteroduplex analysis between λ d*gal* and wild-type λ. The DNA from two types of phage is isolated, denatured, renatured, and viewed through the electron microscope. While the homologous DNA from two sources (phage DNA) will make a perfect double-stranded pair, the non-homologous region (bacterial DNA) will appear as a single-stranded bubble. The branch point of the bubble will define the location, and the loop size will suggest the extent of substitution.

5.4.2 Transfer of the Transduced Material-Formation of Transductant

Once a transducing particle infects a recipient cell, it delivers the genetic material, which now consists of part phage and part bacterial regions. One important prerequisite is that phage should enter into a lysogenic relationship. Experimentally, this could be established by high MOI, and late log phase of the host cells. The bacterial material can then integrate to produce transductants, of which two types can be observed. We shall follow the generation of Gal^+ transductant, though the steps will be the same for Bio^+ types.

Type I transductants are stably *gal*$^+$ and there is no sign of phage DNA. One could postulate that such a transductant will be produced by two crossovers between the recipient *gal*$^-$ and incoming donor *gal*$^+$ region. This will involve simply the substitution of the *gal*$^-$ by *gal*$^+$ and will be scored as the stable change in the phenotype. Type II transductants, however, are unstable and lysogenic to the carrier phage genome. Such a pathway can be explained by the occurrence of a single crossover between the *gal*$^+$ and *gal*$^-$ genes leading to the addition of whole phage genome along with the bacterial *gal*$^+$. As a result, the recipient will now carry two copies of the *gal* gene, both *gal*$^-$ and *gal*$^+$ (partial diploid/heterogenote), and because of the dominance of the *gal*$^+$ allele, the transduced phenotype will be expressed. However, such colonies will be highly unstable as two copies of a gene can facilitate intramolecular recombination, leading to the elimination of the *gal*$^+$ along with λ genome and the recipient will return to its original *gal*$^-$ status (Fig. 5.10).

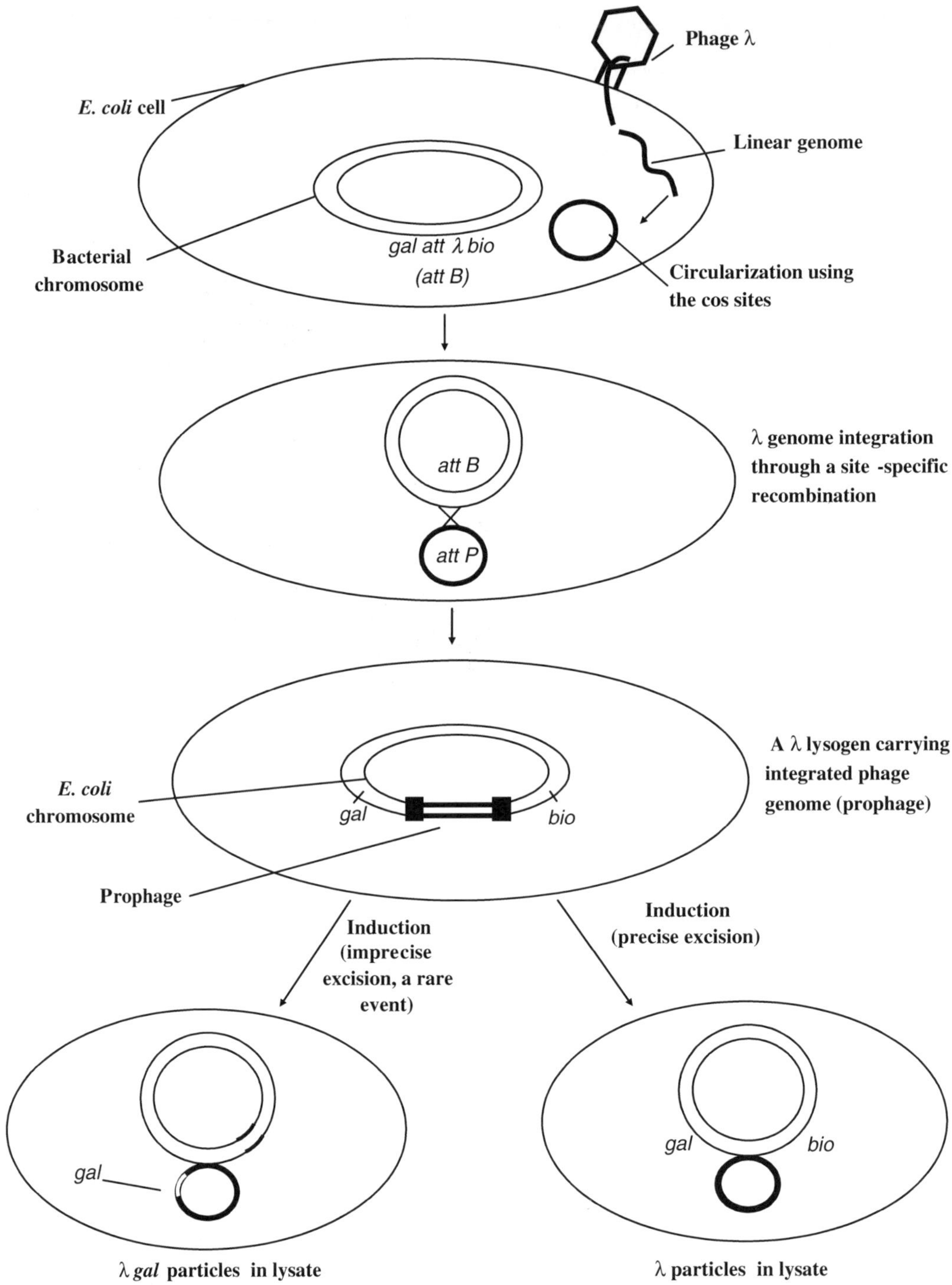

Fig. 5.9 Generation of specialized transducing phage by imprecise excision of λ prophage that results from inclusion of bacterial *gal* genes in phage λ genome

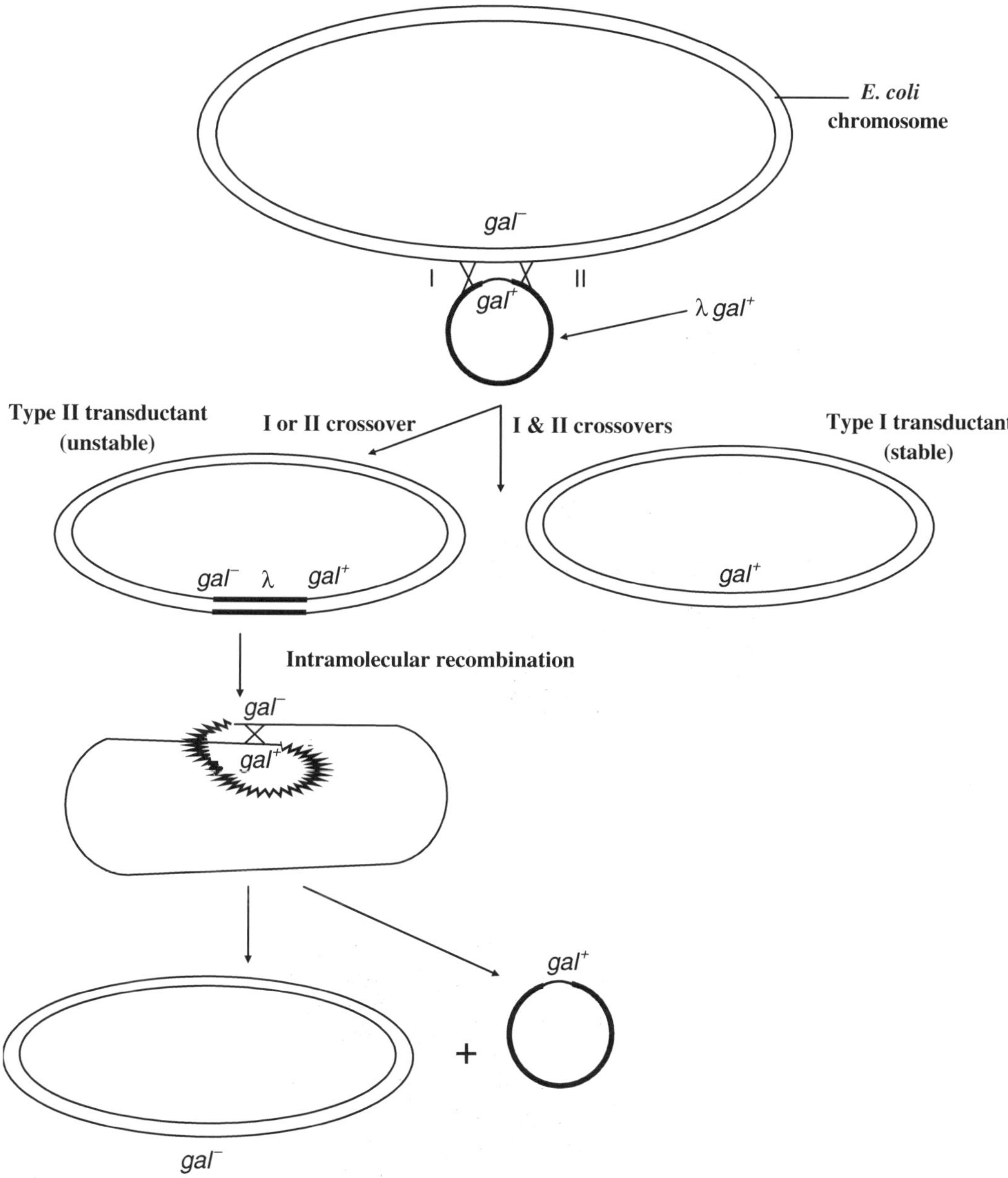

Fig. 5.10 Formation of specialized transductant. This involves the infection of an *E. coli* cell (*gal* $^{-}$) by λ *gal* $^{+}$. The *gal* $^{+}$ gene pairs with the *gal* $^{-}$ of the recipient cell. If two crossovers (I and II) take place, it generates a complete *gal* $^{+}$ transductant (type I) but a single crossover will integrate λ *gal* $^{+}$ genome that leads to an unstable Type II transductant. Two copies of *gal* genes in such transductants are then engaged in intramolecular recombination that results in deletion of λ along with *gal* $^{+}$ gene

At high MOI, when the phage lysate is expected to carry both λ^{+} and λ d*gal*, both genomes may integrate into host genome. First, λ^{+} will integrate using *attB/attP* homology, generating two hybrid sites, *attL* and *attR*. Now, λ d*gal* that carries the *attL* site will utilize this homology to integrate. After these integrations, there will be two tandem copies of λ, one of λ^{+} and another of λ d*gal*, providing conditions for intramolecular recombination. A similar situation arises when λ

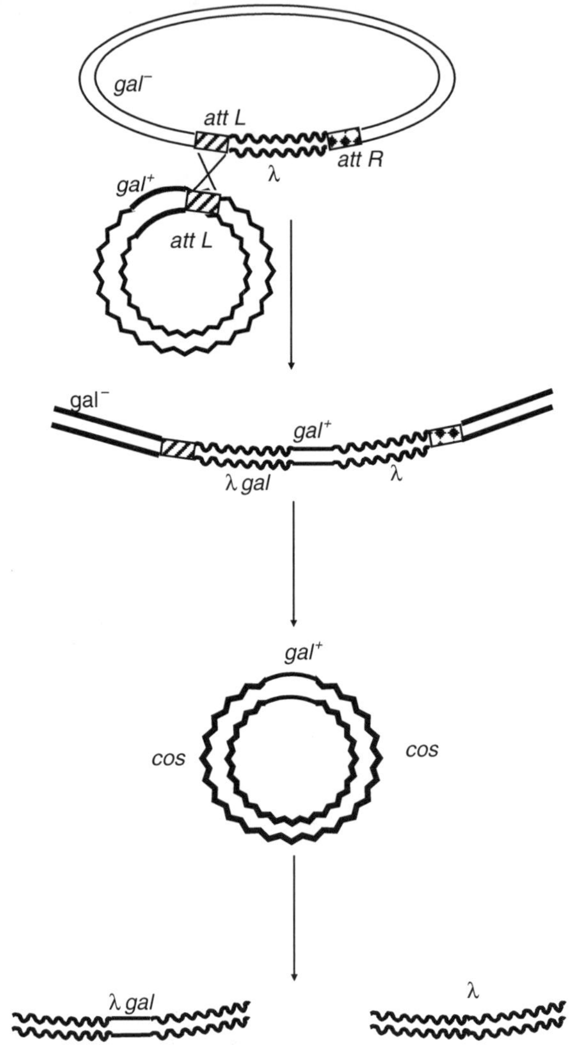

Fig. 5.11 Generation of HFT lysate. When λ *gal* $^{+}$ infects a lambda lysogen its genome integrates using *attL–attL* homology to generate a dilysogen. When induced it will produce two types of progeny λ and λ *gal* $^{+}$ in equal proportions

d*gal* infects a lysogen; a structure so formed is referred to as a dilysogen (Fig. 5.11).

The formation of a dilysogen can be useful because λ d*gal* will be helped in multiplication by λ^{+}, as the latter will provide all the missing functions of the former. Also, the normal packaging of the phage from a dilysogen will produce two types of progeny, λ^{+} and λ d*gal*, at a frequency of $\sim$50 % each. This gives rise to what is known as high frequency transducing lysate (HFT), in comparison to the low frequency transducing (LFT) lysate that is produced by a rare aberrant excision of the phage from a lysogen.

5.5 Transduction in Genetic Analysis

Transduction, like the other two mechanisms of transformation and conjugation, provides a third pathway for gene transfer in bacteria. The relevance of this transfer is translated into genetic exchange that leads to variability, so important for evolution.

Geneticists, however, have also used this mechanism in genetic analysis, wherein the co-transduction frequency between two markers is used as a measure of linkage. Such an analysis

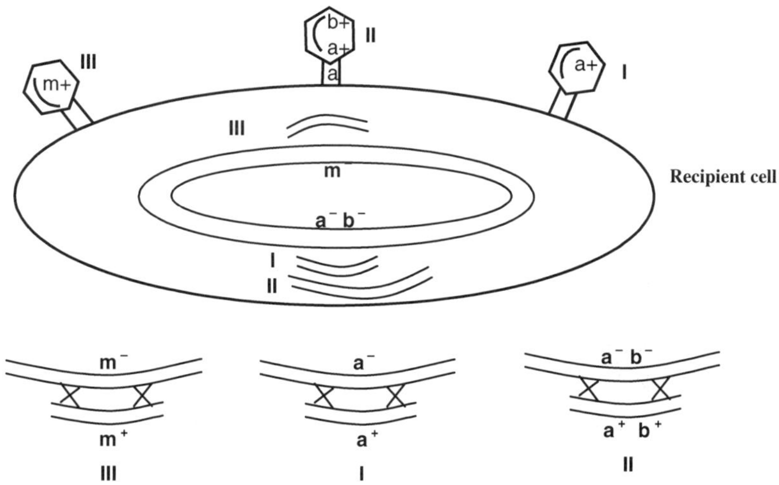

Fig. 5.12 Complete transductant formation—bacterial DNA carried by the transducing particles will be deposited into a recipient cell, where it can undergo crossing over (minimum of two) to replace its counterpart. The event II will take place if two genes are located so close that they are carried on the same DNA fragment. In such a case, a co-transductant can be formed. Co-transductant can also be formed for unlinked genes if the two phages (I and III) simultaneously infect a bacterial cell

has basically employed generalized transduction. The amount of DNA that can be accommodated in a phage head may carry a number of genes. The premise once again is that the closer the two genes, the more is the probability that they are carried on the same piece of DNA to be included in the same phage head. However, transduction for two markers may also take place if a host cell is simultaneously infected by two phages each carrying the gene in question, but singly (Fig. 5.12). Such a situation can be sorted out by analyzing co-transduction at a low MOI, wherein transduction for two unlinked markers would be virtually impossible. When such an exercise is carried out with genes that are known to be close to each other, the co-transduction data do give information on linkage.

Suppose, a^-b^- recipient is transduced with DNA carrying linked a^+b^+ genes, one could now select for b^+ (or a^+), and subsequently classify them to be either a^+ or a^- (or b^+ and b^-). Two types of colonies can be observed: those that are a^+b^+, are due to simultaneous transduction of both, and those that are a^-b^+ will arise due to recombination between them (Note: because of initial selection for b^+, you will not get any combination with b^-, such as a^+b^-).

The co-transduction of two markers can be employed to determine the linkage and thus produce a relative map of the transduced genes. The relationship is the same as described above. In other words, co-transduction frequency is inversely proportional to the distance between the two markers. If for example, we take three markers, co-transduction frequency between the first two or the second and third in a linear sequence will be more than between the first and the third. Many times, however, mere co-transduction frequency may not tell us unambiguously the order of the three genes. For example, if a and b are co-transduced at a frequency of 80 %, b and c at 20 %, and a and c at 18 %, it is difficult to say whether the order is abc or acb, though one can only rule out the order bac. In such a situation, it is best to conduct a pair of reciprocal crosses and determine the frequency of a particular recombinant class (the principle is the same as that discussed under conjugation).

Co-transduction frequency can also be applied to a mathematical formula devised by Wu, to arrive at the distance between the markers. This stands as:

$$F = (1 - d/L)^3$$

where F is the co-transduction frequency, d is the distance in minutes between two markers, and L is the size of the fragment in minutes carried by the phage.

Phage P1-mediated transduction was successfully employed to prepare the fine structure map of *trp* region of *E. coli* by Yanofsky and Lennox in 1959. For this, both the donor and the recipient carried different *trp* mutations, and the two alleles in the host cytoplasm were expected to undergo recombination if located at different sites, which was detected by the selection of trp^+ transductants. This information was then utilized for the construction of the map.

Specialized transducing phages also have a variety of uses. This may range from bringing defined changes in the recipient genome (strain construction), to isolate a gene from the host genome (e.g., λ *gal*), or even creating deletion mutations of the phage, which can be employed for deletion and physical mapping. Many specialized transducing phages have been used as excellent cloning vehicles as well; phage λ and its derivatives being the classical examples.

Further Readings

Ebel-Tsipis J, Fox MS, Botstein D (1972) Generalized transduction by bacteriophage P22 in *Salmonella typhimurium* II. Mechanism of insertion of transducing DNA. J Mol Biol 71:449–460

Fineran PC, Petty NK, Salmond GPC (2009) Transduction: host DNA transfer by bacteriophages. In: Schaechter M (ed) The encyclopaedia of microbiology, 3rd edn. Elsevier, Oxford, UK, pp 666–679

Gingery R, Echols H (1968) Integration, excision, and transducing particle genesis by bateriophage λ. Cold Spring Harb Symp Quant Biol 33:721–728

Ikeda H, Tomizawa J (1965) Transducing fragments in generalized transduction by phage P1. I. Molecular origin of the fragments. J Mol Biol 14:85–109

Lederberg J (1956) Genetic transduction. Am Sci 44(3):264–280

Lennox ES (1955) Transduction of linked genetic characters of the host by phage P1. Virology 1:190–206

Luria SE, Fraser DK, Adams JN, Burrous JW (1958) Lysogenization, transduction, and genetic recombination in bacteria. Cold Spring Harb Symp Quant Biol 23:71–82

Margolin P (1987) Generalized transduction. In: Ingraham JL, Neidhardt FC (eds) *Escherichia coli* and *Salmonella typhimurium*: cellular and molecular biology, 1st edn. ASM Press, Washington DC, p 1154–1168

Masters M (1996) Generalized transduction. In: Neidhardt FC et al (eds) *Escherichia coli* and *Salmonella typhimurium*: cellular and molecular biology, 2nd edn. ASM Press, Washington DC, pp 2421–2441

Miller RV (2001) Environmental bacteriophage–host interactions: factors contribution to natural transduction. Antonie van Leeuwenhoek 79:141–147

Ozeki H, Ikeda H (1968) Transduction mechanisms. Annu Rev Genet 2:245–278

Sato K, Campbell A (1970) Specialized transduction of galactose by lambda phage from a deletion lysogen. Virology 41(3):474–487

Sternberg NL, Maurer R (1991) Bacteriophage-mediated generalized transduction in *Escherichia coli* and *Salmonella typhimurium*. Meth Enzymol 204:18–43

Thompson LC, Costantino N, Court DL (2007) *E. coli* genome manipulation by phage P1 transduction. Curr Proto Mol Biol 79:1.17.1–1.17.8

Trevors JT (1999) Evolution of gene transfer in bacteria. World J Microbiol Biotechnol 15:1–6

Weinstock GM (2002) Transduction in gram-negative bacteria. In: Streips UN, Yasbin RE (eds) Modern microbial genetics, 2nd edn. Wiley- Liss, Inc, p 561–579

Weisberg RA (1996) Specialized transduction. In: Neidhardt FC et al (eds) *Escherichia coli* and *Salmonella typhimurium*: cellular and molecular biology, 2nd edn. ASM Press, Washington DC, pp 2442–2448

Zinder ND (1992) Forty years ago: the discovery of bacterial transduction. Genetics 132:291–294

6 Plasmids: Their Biology and Functions

The genomic constitution of several bacteria may consist of two components, viz., chromosome that carries genes for all essential functions and their regulation, and an extra-chromosomal but autonomous unit, plasmid that was initially thought to carry functions required for its own replication, maintenance, and distribution. However, plasmids should not be considered selfish or promiscuous molecules as many of them have been described to carry genes providing functions of selective advantage. For example, they may code for resistance to several antimicrobial agents, such as antibiotics, heavy metals, or toxins, or for production of antibiotics, pigments, toxins, H_2S, and others, or may provide unusual catabolic capabilities, or endow fertility, etc. They may also induce plant tumors, and other symbiotic and pathogenic responses in plants and animals, to name a few. There may also exist another category, however, where a plasmid may carry some essential genes as well.

Plasmids have thus played a very significant role in bacterial evolution as plasmid-containing cell may have had an adaptive edge over the plasmid-free ones. In recent years, their significance has increased tremendously due to their application in genetic engineering research as a carrier of the foreign molecule (recombinant DNA), and has brought them widespread recognition. A bacterial cell may or may not carry a plasmid or may carry more than one type or copies of a plasmid. Plasmids have often been equated with living organisms by applying the same attributes as is done for viruses. Thus, an "organism is the unit element of a continuous lineage with an individual evolutionary history". This definition fits very well on viruses, plasmids, and certain transposons, all of which have been considered to belong to a family of primitive organisms. The common feature between these units is their replication, maintenance, and dissemination.

6.1 Detection and Nomenclature

Plasmids can be detected by physical as well as genetic means. It was through the latter mode that they were identified to begin with. The detection is based on the phenotype(s) conferred by the presence of a plasmid and its extra-chromosomal inheritance pattern. For example, antibiotic resistance or resistance to other microbiocidal compounds can be detected by growing bacteria on the medium supplemented with appropriate concentrations of such toxic compounds. Similarly, the ability to utilize an unusual compound can be assessed on a medium containing the same. Other plasmid-encoded functions may require specific conditions and vary from phenotype to phenotype (Fig. 6.1). Sometimes, such an association of a character/s with a plasmid is confirmed by effecting plasmid curing, as we shall discuss later in this chapter. Very often, this results in *en bloc* loss of many characters, all of which could be tentatively located on a plasmid. Similarly, transfer of a plasmid encoded function from plasmid-carrying cells to plasmid-free cells

S. Srivastava, *Genetics of Bacteria*, DOI: 10.1007/978-81-322-1090-0_6, © Springer India 2013

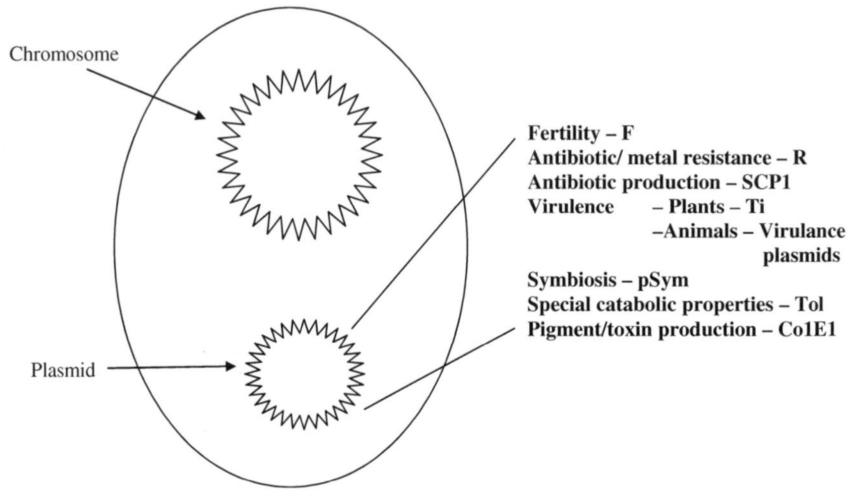

Fig. 6.1 Bacterial cell may carry in addition to its chromosome, an extrachromosomal component called plasmid. Plasmids have been characterized to possess many special properties, some of which are listed along with the commonly known examples

independent of the chromosomal transfer will also suggest the plasmid location of the concerned function(s). Physically, plasmids can be detected by agarose gel electrophoresis of either the crude cell free lysate or a purified preparation, for which a large number of protocols are available now. The separation is based on differences in the gel mobility of comparatively smaller, covalently closed circular (ccc) DNA molecules such as the majority of the plasmids, and the linear fragments of the chromosome generated during the preparation (Fig. 6.2). Another commonly used method is the separation on CsCl–EtBr density gradients based on the differential stacking of EtBr, and resultant changes in the density of ccc and linear DNA. Very large plasmids can also be detected by pulse field gel electrophoresis (PFGE).

Plasmid nomenclature is generally derived from the functions/genes they carry. For example, R plasmids carry genes for resistance to several antibiotics, ColE1 carries the gene for bacteriocin, colicin E1, Tol plasmid reflects the presence of genes for toluene degradation, Ti plasmid of *Agrobacterium tumefaciens* harbors the property to induce tumors in some plants, and F plasmid is responsible for conferring fertility to *E. coli* cell possessing the same, to name a few.

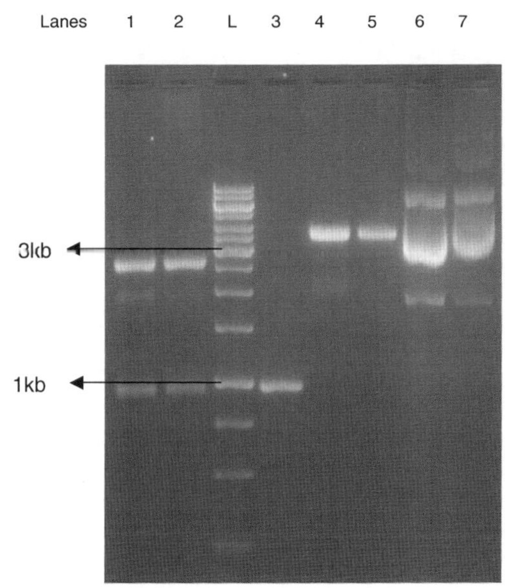

Fig. 6.2 Detection of plasmids by agarose gel electrophoresis. A plasmid preparation is run on an agarose gel and covalently closed circular (ccc) DNA molecules (plasmid) can be identified by EtBr staining. Plasmids may often show the open circular (oc) and linear forms (l) due to mechanical breakage (lanes 6 & 7). The size of the plasmid can be determined by linearizing/digesting the same by a restriction enzyme (lanes 1&2 and 4&5) and running a DNA size ladder (L) along it

Based on a more standardized practice, a plasmid is labeled by the prefix "p", followed by capital letters that suggest the plasmid function

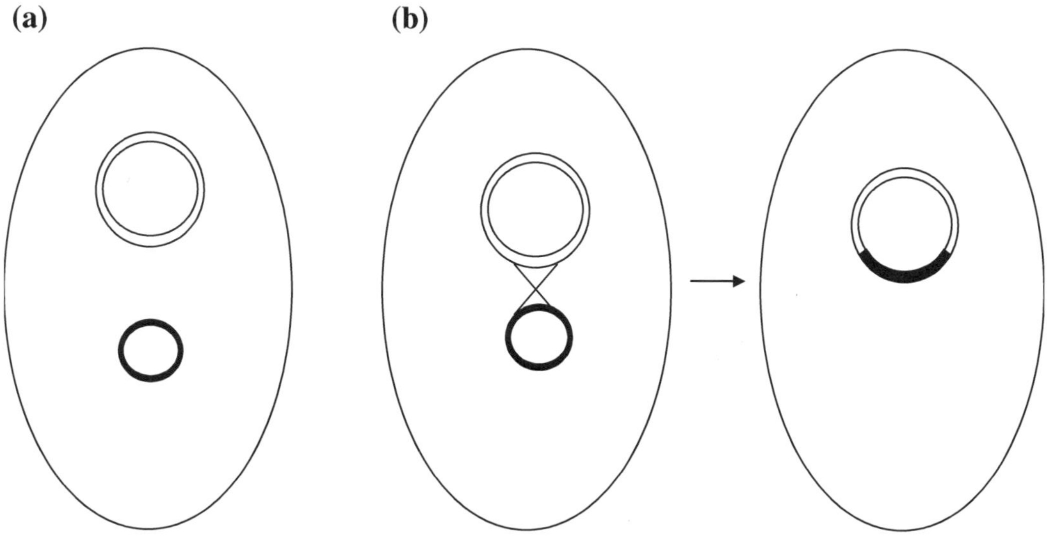

Fig. 6.3 Plasmid may exist as free autonomous structure (**a**) or may integrate into the bacterial chromosome (**b**). In the latter condition, the plasmid is referred to as episome

(R, F, Tol, ColE1) or initials of the person or persons who isolated/constructed it (e.g., pBR represents the names of Boliver and Rodriguez), or the strain from which the plasmid has been derived (pSCP—*Streptomyces coelicolor* plasmid). In fact a large variety may be seen in the way a plasmid may be labeled and once a new plasmid is reported, the author is required to provide a complete description of its important features.

6.2 Plasmid Organization

Most plasmids are covalently closed circular (ccc) pieces of DNA, varying in size from a few thousand to hundreds of thousands of base pairs. These occur in a cell in a negatively supercoiled state. Although identified as free autonomous genetic units, some plasmids have the ability to integrate into the bacterial chromosome (recall F plasmid). Such plasmids are referred to as episomes (Fig. 6.3). Interestingly, members of the genus *Streptomyces* and *Borrelia burgdorferi* possess linear plasmids. Some very large plasmids have often been referred to as mini-chromosomes or mega-plasmids. The distinctive autonomously replicating nature suggests that a plasmid can be basically equated with a replicon and must code for some or several functions required for its replication. To be autonomous, they must have at least one origin of replication (*ori*), a *cis* site, and must also code for a plasmid-specific protein required for *ori* recognition and thus replication initiation. The *ori* site is generally located within a several hundred base pair segment, a feature that has been thoroughly exploited in manipulating the plasmid size during vector construction. The *ori* thus serves as the site where plasmid and the host (bacterial) proteins are engaged in replication initiation. The plasmid-specific replication initiation protein Rep is coded by *rep* gene carried by the plasmid. The Rep protein is generally capable of acting in *trans,* though in some plasmids, it preferentially acts at its own *ori*, i.e., in *cis*.

In some linear plasmids, as in *B. burgdorferi*, the ends may bear telomere-like sequences, that is, the ends of the DNA may be repeated and may even be covalently joined to each other. Similarly, in *Streptomyces,* plasmid ends may be linked with a protein (Fig. 6.4a, b). Both these features have evolved to take care of the problem of the ends priming, and they may share these structural features with some eukaryotic viruses.

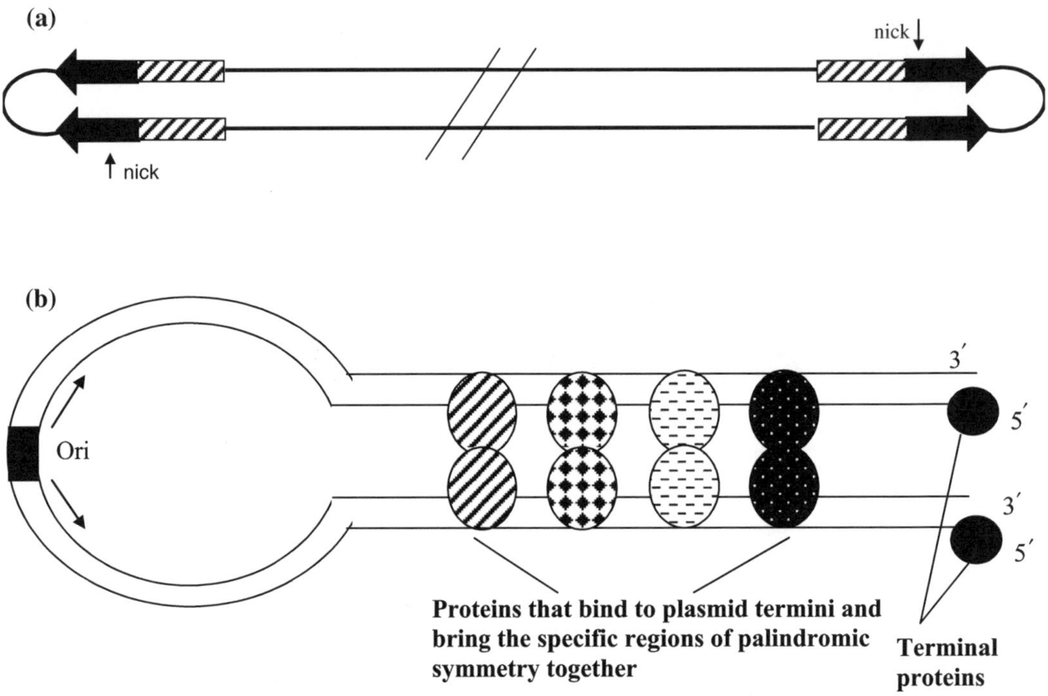

Fig. 6.4 **a** Linear plasmid of *Borrelia* containing terminal hairpin telomeric structures composed of inverted repeats, harboring the nick sites involved in DNA replication. **b** Linear plasmids of *Streptomyces* species. Here the 5′ ends are bound to a protein that not only protects the ends but also completes the replication of both termini. Several other proteins bound to specific regions of palindrome symmetry keep the two linear arms together. *Ori* is located near the *center* of the plasmid where bidirectional replication is initiated

Besides the replication function, plasmids may also code for a variety of functions, as we discussed earlier.

6.3 Replication

Replication is one property of the plasmids that has important bearing on their other functions, as we shall discuss throughout this chapter.

Besides the *ori,* often called *oriV*, and the Rep, the replication initiation protein, plasmids in general depend upon the chromosomal replication functions, such as for the enzymes like, DNA polymerases, helicase, primase, ligase, and others. The circular plasmids replicate by one of the two basic mechanisms: (i) theta mode for both unidirectional and bi-directional pathways and (ii) rolling circle mode.

6.3.1 Theta Mode

In this pathway, replication begins with DNA unwinding at *ori* site creating a structure, where the whole replicational machinery assembles. The structure resembles the greek letter theta (θ) from which the name has been derived. The RNA primer that initiates the process is synthesized and then it is extended by the addition of deoxyribonucleotides. The process of replication may proceed in one (unidirectional) or both the directions (bi-directional). In the former, a single replication fork moves around the circle until it returns to its point of origin. In the other, two replication forks travel in opposite directions beginning at *ori,* until they meet at some point on the other side of the molecule (Fig. 6.4). This is the most common mode of DNA replication, followed by plasmids such as ColE1, F, P1, and RK2. Readers may recall that

E. coli chromosome also follows a similar pathway during its replication.

6.3.2 Rolling Circle Mode

This type of replication requires the nicking of one DNA strand at the *ori* region. While the 3′ OH end at the nick site serves as the primer to which fresh deoxyribonucleotides can be added leading to the displacement of the original strand. The other strand then acts as the template. One can then envisage a template strand continuously rolling as it replicates, dishing out several copies of single-stranded DNA. These single strands are then converted into double-stranded structures to complete the replication (Fig. 6.5). Plasmids using rolling circle mode include, pUB110 and pC194 of *Staphylococcus aureus*, pIJ101 from *Streptomyces lividans*, and F plasmid of *E. coli* at the time of its conjugational transfer. Single-stranded plasmids in many Gram-positive bacteria arise from defective rolling circle replication.

The dependence on host's enzymes for plasmid replication varies from plasmid to plasmid. While some may use DNA polymerase III for chain elongation (F), others may need DNA polymerase I to initiate and polymerase III for elongation subsequently (ColE1). Generally speaking, most plasmids use host encoded products that include DnaB (a prepriming protein), DnaE (coding for catalytically active subunit of DNA polymerase III), and DnaG (primase) as well as ligase, other enzymes and protein factors.

Linear plasmids have been detected in fungi, whereas it was generally believed that prokaryotes contain only circular plasmids. This myth has been broken as many bacteria, such as *B. burgdorferi* and *Streptomyces* species are known to contain not only linear plasmids, but also interestingly the linear chromosomes. Linear plasmids have to maintain their ends intact as these ends may be highly recombinogenic. This may also lead to the problem of priming at the ends that will shorten the plasmids and make them highly unstable. Eukaryotic chromosomes have evolved the highly repetitive telomeric structures to overcome such problems. *B. burgdorferi* linear plasmids have their ends in a covalently closed form and *Streptomyces* plasmids have adopted the strategy of being blocked at their 5′ ends by a protein, like many viruses. In both the strategies, the ends are protected and plasmids are maintained stably.

The fact that plasmids' origin of replication and other accessory replication functions are confined to a small region, has been thoroughly exploited in plasmid manipulations for their usage as vectors in r-DNA technology. Thus, a suitable plasmid can be constructed by removing large stretches of the original DNA, without jeopardizing its replication, and therefore its maintenance and stability. Most plasmids have been described to possess a single *ori* but large plasmids with multiple origins are frequently encountered. For example, F plasmid has two *ori* sites (with an additional inactive *ori*) and plasmid R6K has three *ori* contained within a ~4.0 kb segment.

6.4 Copy Number Control

A plasmid-containing cell may carry a specific number of each plasmid, referred to as its copy number. This number defines the number of copies of a plasmid that a newly born cell possesses immediately after the cell division. This number may vary from as small as one to a few hundreds. Accordingly, the former is known as low copy number, and the latter the high copy number plasmids (Table 6.1).

The characteristic copy number is understandably regulated by replication, in which the *ori* region of the plasmid plays a significant role. The mechanism of regulation employed by two categories of plasmids differ. While the high copy number plasmids only need to inhibit the replication initiation once they reach their characteristic number, the low copy number plasmids exert stricter regulation, not only on replication but also on their segregation. This has led to another way of classifying plasmids into relaxed (high copy number) and stringent (low copy number) types. Different plasmids

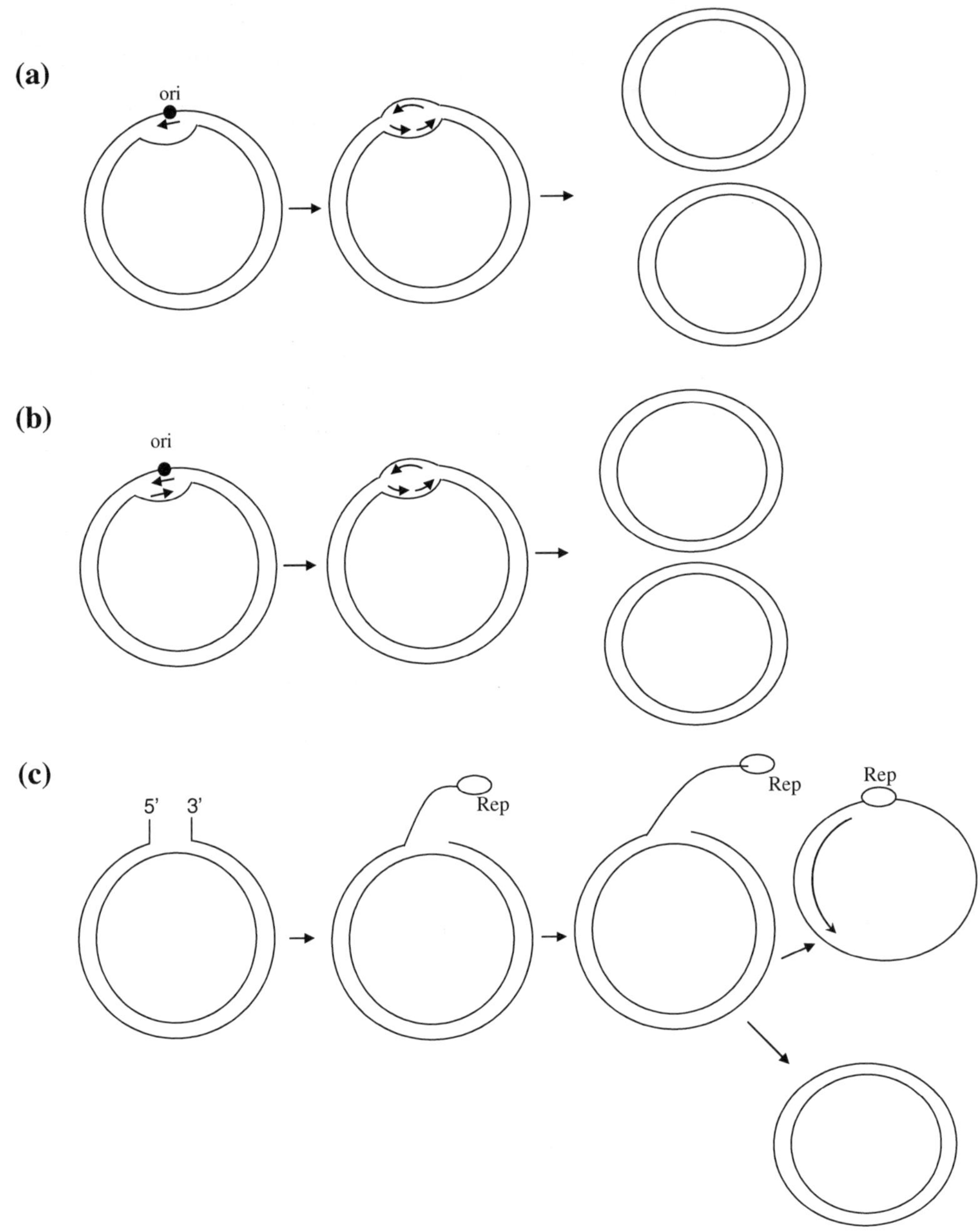

Fig. 6.5 Some common mechanisms of plasmid replication. (**a**) Theta mode unidirectional, where replication fork goes back to origin to complete replication, (**b**) Bidirectional, where two replication forks start at *ori* but in opposite orientation and meet somewhere opposite to the origin to terminate replication, (**c**) Rolling circle replication, begins by adding dNTPs at the 3′ nicked end which continually displaces the 5′ nicked end until it replicates the template strand. The two ends of displaced strand are rejoined and the complementary strand is synthesized from RNA primers that initiate at a unique site

have evolved different mechanisms to control their copy number. Two important inputs are the *ori* sequence and the control exerted on it either by an RNA (a counterscript), or protein, or both. We shall discuss these by citing specific examples.

Table 6.1 Size, nature, and copy number of some plasmids

Plasmid	Nature	Approximate size (kb)	Approximate copy number
F	Circular	100	1–2
P 1	Circular	94	1–2
R K2	Circular	56	5–8
R6 K	Circular	38	15–30
ColE1	Circular	6.6	15–30
pBR322	Circular	4.36	15–20
pUC	Circular	2.68	>100
pSC101	Circular	9.26	5
pACYC184	Circular	4.24	18–20
pUB110	Circular	2.3	20–50
pSymA	Circular	1354	2–3
Ti	Circular	205	1–2
pIJ101	Circular	8.9	40–300
lp25	Linear	24.2	1–2
SCP1	Linear	350	4
pB toxis	Circular	128	1–2

6.4.1 Col Plasmids

The prototype ColE1 plasmid produces colicin E1. This and related plasmids are small and are found in several copies per cell. The replication of ColE1 and other similar plasmids, such as, RSF1010, pMB1, CloDF13 draw upon a number of host functions in the form of DNA Polymerase I and III, RNA Polymerase, and products of *dnaB*, *C*, *G*, and *Z* genes. As these plasmids do not require *de novo* protein synthesis, their replication remains largely unaffected by the protein synthesis inhibitors, such as chloramphenicol. This may lead to preferential amplification of the plasmid and provides an important attribute to them.

DNA replication starts at *ori* region and is initiated by transcription of an RNA pre-primer, ~500 bp upstream of the *ori*. The pre-primer RNA hybridizes with the corresponding DNA strand, and the DNA–RNA hybrid is processed by RNaseH to provide the primer function. To this primer, dNTPs are added during replication. The plasmid replication control, and thus the copy number is brought about by the synthesis of an RNA-I, an anti-sense RNA, or a counter-script. This, ~100 nucleotides long RNA is transcribed in the opposite direction (from the other DNA strand) to that of RNA primer. Thus, RNA-I now hybridizes with the pre-primer that cannot be recognized by RNaseH, preventing its processing. The negative control exerted by RNA-I results in the blockage of primer generation and therefore the replication. The mechanism can be further supplemented by the activity of the *rop* (repressor of primer) gene, also sometimes referred as *rom* (RNA on modulator). Rop, a 63-amino-acids protein accelerates and stabilizes the interaction between RNA-I and pre-primer, thus exerting further control (Fig. 6.6). The sequence of RNA-I and the pre-primer are such that they can form stem-loop structures having potential to base pair between them. Additional copies of RNA-I gene provided to a cell results in further decrease in the copy number, confirming this to be the principal mechanism. RNA-I–primer interaction also seems to be involved in plasmid incompatibility.

6.4.2 IncFII Plasmids

Plasmids belonging to this group (R1, R6-5, R100) are self-transmissible, conjugative plasmids found in Enterobacteriaceae. They are low copy number, fairly large (~100 kb) plasmids,

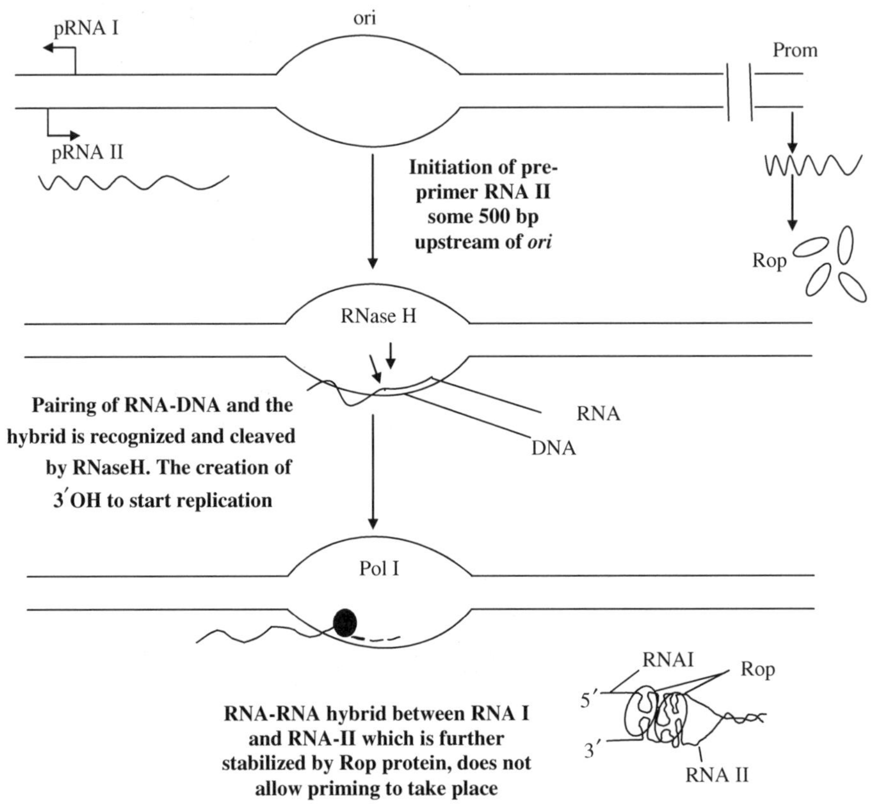

Fig. 6.6 Replication of ColE1 type plasmid and its copy number control

requiring *E. coli dnaB*, *C*, *E*, *F*, and *G* gene products for replication.

Plasmid copy number is mediated by a complex mechanism. The gene *repA* codes for the positive regulatory protein required for replication initiation and works only in *cis*. This gene is transcribed from two promoters, *repAp* and *copBp*. The latter transcript produces a fused product (CopB–RepA), whose 11 kDa CopB product negatively regulates *repAp* transcription. Thus, under normal circumstances, RepA is almost exclusively produced from the *copB* transcript. This may create an impression that the *repAp* is superfluous but the answer is no. In fact, this promoter can be effectively utilized when the plasmid gets transferred to a plasmid-free cell, where quick replication will be critical to establish the plasmid in the new environment.

Superimposed on this mechanism is another mechanism important from the point of view of copy number control. The *copA* gene codes for a ~90 nts RNA (RNA-I) that acts as a counter-script of a part of *copB* product, and the two get involved in a specific interaction. The part that acts as the target of RNA-I action is *copT*, a region on the *copB–repA* transcript. The basis of this interaction is a high degree of secondary structure that produces stem-loop formations, matching with similar structures on *copA* RNA by complementary base pairing. The *copA* product thus functions by interfering with the translation of *repA* mRNA. An additional input is provided in the form of Tap (translation activator peptide), a 24 amino acid peptide, coded by an ORF located between the *copB* and *repA* reading frames. While the translation of *tap* is required for expression of *repA*, *copA* can inhibit the synthesis of Tap. Thus, the *copA* anti-sense RNA indirectly controls the expression of *repA* by exerting its regulation on *tap* translation (Fig. 6.7).

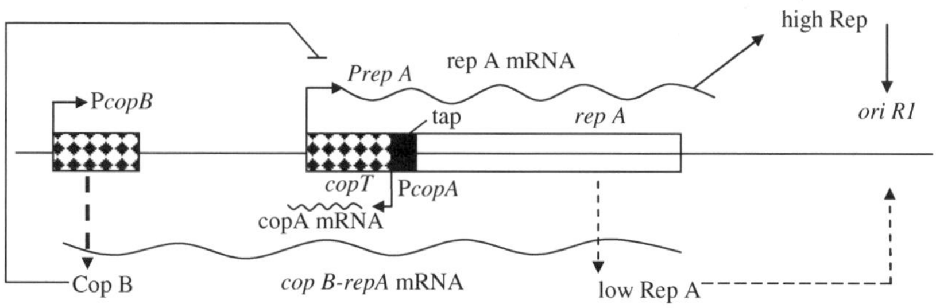

Fig. 6.7 Copy number control in plasmids of IncF II group (R1). Rep A is the positive regulator of *ori*R and is synthesized from two transcripts *repA* and *cop B-repA*. The CopB product is a negative regulator of P*repA* and thus all RepA is synthesized from a weak P*copB*. An anti-transcript *copA* interacts with the *copT* region of *repA* transcript. This interaction involves the loops formed in the secondary structure of two RNAs with a potential to pair. This interaction also inhibits the translation of *tap* and, therefore, of *repA*

Different IncFII plasmids carry a highly conserved inverted repeat sequence that in transcribed RNA could fold back to form a stem-loop structure, hiding the initiating GUG codon and the Shine–Dalgarno region from the ribosome. Tap is proposed to modulate this structure so that the nearby ribosome-binding site for *repA* translation becomes available. The function of CopB is to provide a switch between a low and a high copy number state. The *copA*, on the other hand, helps in maintaining a constant copy number under both the situations.

6.4.3 Control by Iteron Binding

Several plasmids, such as F, pSC101, R6 K, Rts1, RK2, and phage P1 (which under lysogenic state or prophage exists in the form of a free ccc DNA) regulate their copy number by this mechanism. The replication of P1 plasmid requires about 1.5 kb region consisting of: (i) *oriR* of ~245 bp, (ii) the *repA* gene coding for 286 amino acid replication initiation protein, and (iii) *incA*, a 285 bp stretch that controls replication.

The *oriR* carries a consensus sequence for *E. coli* DnaA protein binding and bears homology with other low copy number plasmids. Replication of P1 depends on *dnaA* function suggesting that DnaA binds to the conserved sequence. The binding site in the *ori* region lies next to an A–T rich region that includes a repeat of 5 GATC sequences. In addition, there are five 19 bp repeat sequences referred as iterons. These closely spaced iterons lined up in the same orientation constitute the *incC*. Nine more iterons located downstream of *repA* comprise a region called as *incA* (Fig. 6.8).

The protein RepA is required to initiate unidirectional replication at *oriR*. RepA is autoregulated in that it represses its own transcription. The specific placement of iterons on the either side of *repA* suggests that RepA could bind to both *oriR* and *incA*, in a concentration-dependent manner. Thus, the binding of RepA to *incA* will make it unavailable for any DNA synthesis activity at *oriR* and will also lead to autorepression of its own synthesis from *repA*. Both *oriR* and *incA* compete for RepA, a feature shared with the plasmid pSC101. RepA, therefore, has a dual role: it positively regulates the initiation of DNA synthesis, and negatively controls the expression of *repA* gene. Many investigators have proposed a handcuffing model based on the concentration of RepA as well as the plasmid (essentially the number of iteron sequences). The RepA protein has been implicated in linking the two plasmids by binding to their iteron sequences, thereby preventing any further replication. At low RepA concentration and small plasmid number this is not likely to happen (Fig. 6.9). Another well-known example of this type of the control mechanism is found in F plasmid (see Fig. 6.8).

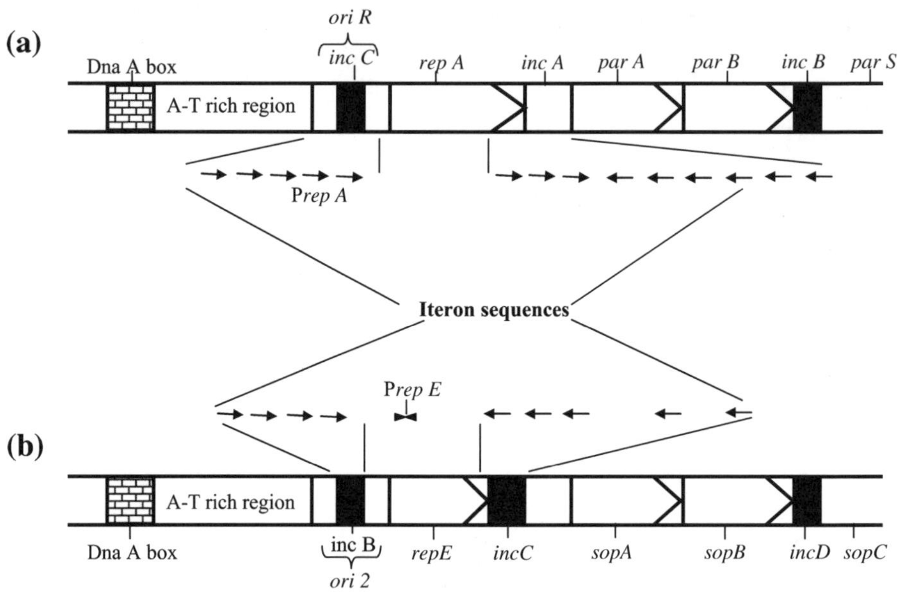

Fig. 6.8 Replication region and the copy number regulatory components of plasmids P1 (**a**) and F (**b**). Note the presence of iteron sequences and the autoregulated initiator proteins RepA/RepE and DnaA binding region. RepA/RepE are also involved in handcuffing the plasmid copies resulting in the further control of copy number

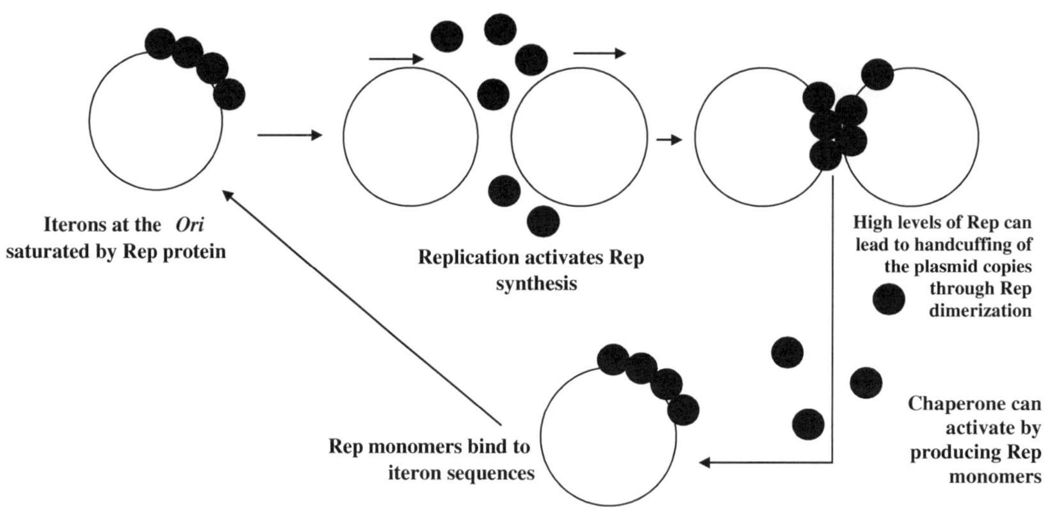

Fig. 6.9 Copy number control through Rep dimerization (inactive and thus cannot bind to iteron sequences) that may lead to handcuffing between plasmid copies and inhibition of replication. The chaperone induces monomerization of Rep (active Rep) that can bind to *ori*/iteron and initiate replication

6.4.4 Other Mechanisms

A number of plasmids, originally discovered in Gram-positive bacteria but subsequently also found representation amongst Gram-negative members and other prokaryotes, employ a rolling circle replication. In this mode, an initiator protein, Rep creates a nick on the leading strand at *ori* region. The nick provides an entry site for the addition of deoxyribonucleotides and this function mediated by DNA Polymerase leads to extension of the strand at the 3′ OH end, with simultaneous unrolling of the template strand. The lagging strand is synthesized from a separate origin that becomes available only on the single-stranded form, using host enzymes and other required factors.

The model plasmid pT181 described from *Staphylococcus aureus* is a small 4.4 kb plasmid carrying *tet*^{r} gene. It has an *ori* region of ~108 nt that partly overlaps with a region of *repC* gene coding for amino terminal portion of RepC. RepC is a positively regulating initiator protein. Its binding to *ori* is facilitated by DNA bending and it functions in the form of a dimer. After each initiation the dimer is inactivated by modifying one of the RepC, resulting in a non-functional heterodimer. The plasmid also carries a region *cmp*, and modification of this region can lead to a sharp drop in copy number (as much as 20-fold). The *cmp* region has been proposed to act as an enhancer and stimulate RepC utilization. Plasmid pT181 also carries a *palA* region. The *palA*$^-$ plasmids are highly unstable and have 10-fold decrease in copy number. The *palA* function is dependent on its orientation to *ori.* A similar region *par* has been described in *B. subtilis* plasmid, pSL11. The copy number control in pT181 and other related plasmids of this group is exercised by a counter-transcript. This anti-sense RNA interacts with leader region of Rep-mRNA causing premature termination. The latter is based on the refolding of Rep-mRNA in a stem-loop structure signaling termination. Many such plasmids may code for more than one counter-transcript.

The plasmid, pLS1 of streptococci is also regulated by an anti-sense RNA. There is, however, another superimposed mechanism consisting of a transcriptional repressor, CopG. This protein binds to the promoter region of *rep* and blocks its transcription.

There are plasmids which can replicate in a variety of hosts. Such plasmids may possess multiple controls. Plasmid RSF1010, an IncQ plasmid, can replicate in several Gram-negative bacteria and codes for three proteins, RepA, RepB, and RepC. While RepC is the positive regulating initiator protein that binds at *oriV*, RepA functions as DNA-dependent ATPase. RepB is a primase that is produced in two forms: one functional 38 kDa product and another larger 70 kDa RepB* of unknown function. These genes are transcribed from p1 promoter, with another promoter lying between *repB* and *repA*, *repC* that can transcribe the latter two independently. Each promoter is regulated by a negative regulatory protein with the end result of controlling the levels of RepC.

6.5 Plasmid Stability and Maintenance

Plasmids are generally described as the extra-chromosomal elements that carry genes for secondary functions. Although several exceptions to this latter part of the description are available now, this may simply mean that cell can lose a plasmid without jeopardizing its own survival. However, several plasmids are known to be highly stable in their host cell cytoplasm, and must have evolved mechanisms to ensure their maintenance. These may include several pathways such as those leading to proper segregation/partitioning, maintenance of the plasmids, especially low copy number ones in a monomeric form, and mechanisms evolved to resist curing. We shall first discuss some general aspects of plasmid maintenance, the fallout of which can be seen as plasmid curing.

If the plasmids do not have a partitioning system, one could calculate as to how often the cells would be cured of a plasmid using a combinatorial probability. Let us assume that the copy number of a plasmid is 4. Each cell will carry eight copies just before the cell division, and the two resulting daughter cells are expected to receive four copies each. However, plasmids are seldom distributed equally to the daughter cells, and the two are likely to receive unequal numbers. Since each plasmid can be distributed to either of the cells, the chance that one cell will get all and another none can be determined on the basis of probability distribution. For example, if the probability that the first plasmid will go to one cell is ½, the probability that the second copy of the plasmid will go into the same cell will be ½ × ½ or ¼, and the same trend will follow for other copies as well. Thus, if we look at the original number the probability that all eight plasmids will go into one cell will be $(^1/_2)^8$. Since this will lead to curing of one of the two cells, the overall frequency of curing can be derived as $2 \times (^1/_2)^8$ or $^1/_{128}$ of the cells at each cell division. In general, the frequency of curing for a plasmid with copy number n will be estimated from the relationship: $2 \times (^1/_2)^{2n}$.

This frequency will be applicable if the plasmid distribution to daughter cells is absolutely random. If, however, the observed frequency is lower, it would suggest the existence of a partitioning mechanism. The curing of a high copy number plasmid due to random distribution may still be difficult, but this may prove deleterious for low copy number plasmids. For example, For P1, and F which have only one copy per cell, this may mean that half the number of the cells may be cured $[2 \times (^1/_2)^{2n}]$ at every cell division. Since this seldom occurs, it suggests that specific mechanism/(s) may operate to ensure proper distribution of plasmids.

6.5.1 Plasmid Curing

Although several plasmids are highly stable, there are many that can be lost spontaneously or can be cured. Historically, the association between a plasmid and its encoded character(s) was determined by effecting plasmid curing. Under such a situation, loss of one or many characters could be observed without changing the chromosome-based characters. Subsequently, such a loss could also be correlated with the physical loss of the plasmid. Such a loss could also not be reverted by any known mutagen.

While several chemical compounds have been tested as curing agents, acridine dyes have been the most popular one. These compounds are known mutagens because of their ability to stack themselves in between the DNA bases (both the plasmids and chromosome), inhibiting replication. However, at a relatively lower concentration, plasmid replication is disrupted but chromosomal replication may go on unhindered. At the time of cell division, this may lead to segregational loss of the plasmid from one of the cells, and thus curing. Plasmid curing is routinely sought for by the bacterial geneticists so as to determine the location of the gene(s) of interest.

6.5.2 Plasmid Partitioning

As discussed earlier, many low copy number plasmids need to operate a partitioning (*par*) system, so as to properly distribute their copies in daughter cells. DNA segregation during mitosis in eukaryotes, is a dynamic process, ascribed to the mitotic spindle and is assisted by a large number of proteins. In contrast, the mechanism of prokaryotic DNA segregation remained largely obscure with the most acceptable model involved the coupling of the replicating DNA to the elongating cellular membrane at the time of cell division as a means of segregation. Evidence, however, kept mounting that stable maintenance of bacterial plasmids, especially the low copy number ones, as well as of the chromosome, is brought about by an efficient partitioning (*par*) system; though other mechanisms may also operate.

Bacterial low copy number plasmids have served as useful model systems to study DNA segregation after replication. This is based on the ease with which plasmids can be studied and manipulated. All studied plasmids carry *par* loci or module that functions by determining the intracellular position of their target replicon and actively segregating the plasmid copies on either side of the cellular septal plane. Thus, after cell division, plasmids are faithfully segregated to the daughter cells.

In general, *par* loci consist of (i) one or more centromere-like *cis* site(s) on the plasmid DNA, (ii) a protein (adapter) that binds to this site, and also interacts with NTPase, and (iii) a protein that acts as ATPase/GTPase (NTPase) and provides the energy or motor function. The adapter proteins, therefore, not only function as a link between the NTPases and the centromeric DNA, but also serve an important regulatory role. All these are required for proper partitioning process, which may vary in molecular details. Based on the nature of their NTPase, the *par* systems can be classified into three types (Fig. 6.10).

The type I partitioning loci, of which P1 and F are well-documented members, consist of *parA*/*sopA*, encoding a Walker-type ATPase with homology to the oscillating MinD protein (a cell division inhibitor protein involved in septal placement), and *parB*/*sopB* whose product binds specifically and co-operatively to a centromere-like site, *parS*/*sopC*. The A protein autoregulates the expression of the *par* genes by binding to its own promoter and also performs the function of correct positioning and, therefore, the faithful segregation of plasmid copies. The B protein by binding to the centromere-like site forms a nucleoprotein complex, essential for partitioning. This complex is also known to mediate the pairing of P1 centromere-like sites. The centromere-like *cis*-acting site, *parS*/*sopC* consists of different numbers of direct or inverted repeats (iterons). In *parS*, this represents an AT-rich region with a 13 bp perfect inverted repeat and in *sopC*, it covers 12, 43 bp direct repeats downstream of *sopB*; these serve as the sites for binding of ParB/SopB. The *parA* and *sopA*, as well as *parB* and *sopB* from two different plasmids share a lot of homology. ParA/SopA are P loop cytoskeletal ATPases, of which two different kinds are known. One class reported only from the plasmids is the one that possesses N-terminal DNA-binding helix-turn-helix (HTH) motif, also called as large ATPase, and the second class, reported both from some plasmids and chromosome, referred to as small, lacks such motifs (see Fig. 6.10). Moreover, ParAs can also bind to DNA co-operatively and non-specifically, which goes in favor of their nucleoid association. It has been shown that ParAs can form dimers and its co-operative association with non-specific DNA only occurs with ATP-bound dimers, leading to the formation of ParA filaments. Such a growing ParA filament makes contact with a plasmid copy via ParB–ParC complex, which in turn stimulates the ATPase activity of the former. This reaction converts ParA–ATP to its ADP form and is released from the DNA. At this stage, the plasmid may also detach or it may remain attached to depolymerizing ParA. A similar polymerization-plasmid contact-depolymerization can eventually take place with a plasmid copy in the opposite direction as well. Thus, ParA is able to provide a pulling force to partition the plasmid copies into two cellular halves. Free ParA-ADP is rejuvenated to ParA–ATP, and the cycle can be repeated (Fig. 6.11).

As shown in Fig. 6.10, the plasmid, pB171 contains two centromere-like sites. This and few others have been described to carry double *par* locus. These called, *par1* and *par2* are two adjacent and oppositely oriented *par* loci of different types. The *par1* encodes an actin-like ATPase (ParM), as described below for type II system, *par2* codes for oscillating ATPase (ParA). Both share a common *cis*-acting region, *parC1*, located between them. While ParB binds to 6-bp direct repeats within *parC1* and *parC2*, ParR recognizes two 10-bp direct repeats in *parC1* located upstream to *parM*. Both ParB and ParR transcriptionally regulate the *par2* and *par1*, respectively.

The second group of partitioning system (type II) exemplified by plasmid R1 of *E. coli* is organizationally very similar to type I. In R1 *par*, centromere-like site *parC* (for centromere) consists of two sets of 5 direct 11 bp repeats or iterons flanking the core promoter. Binding of ParR (for repressor) to *parC* forms a nucleoprotein complex required for pairing of plasmid copies, thereby facilitating their partitioning, and also for autoregulating the expression of *par* genes. The *parM* (for motor function) codes for an ATPase that has homology with eukaryotic actin, like F-actin. ParM filament formation is subjected to two levels of regulation. First, ParM polymerization takes place in the presence of both ParR and *parC;* and second, the ATPase activity of ParM confers dynamic turnover to

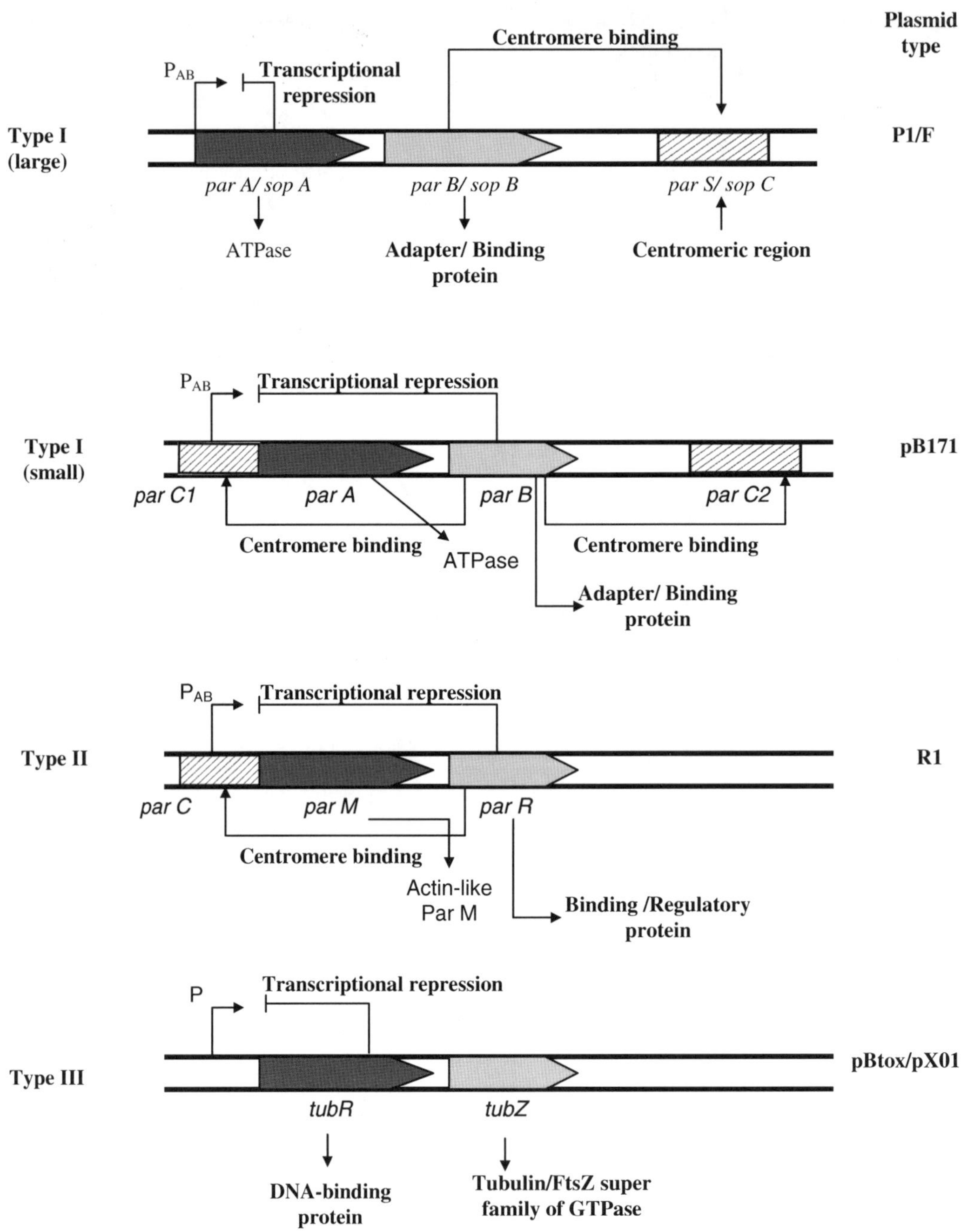

Fig. 6.10 Organization of *par* locus in some representative plasmids along with their broad functions

ParM filamentation only when it interacts with ParR/*parC* complex. Such filaments are also formed only in those cells that contain at least two plasmid foci (copies), suggesting that the plasmid replication is essential. The model for R1 partitioning system thus, envisages that the nucleoprotein complex at *parC*, consisting of ParR and the paired plasmid replicates serve as a nucleation point for ParM polymerization, which in turn acts to push the plasmid copies apart.

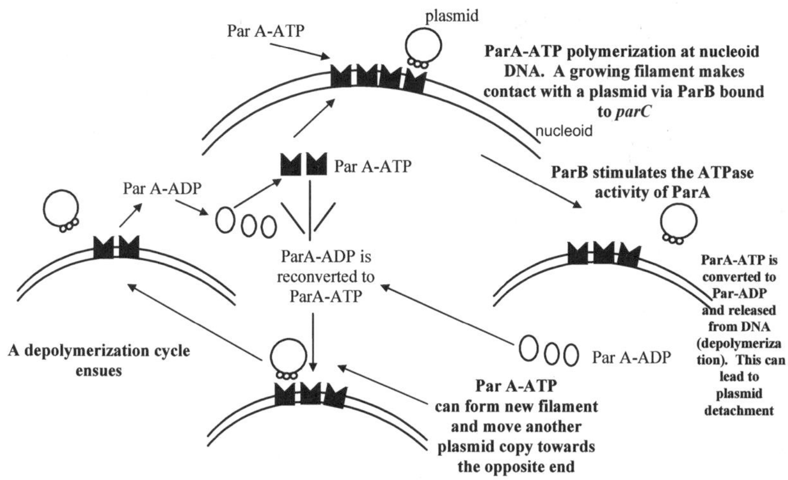

Fig. 6.11 ParA-based plasmid partitioning system (Type I) in which a polymerization–depolymerization cycle deposits the plasmid copies into different cellular halves

The plasmid DNA copies appear to be carried on the tip of the growing ParM filament and pushed in the opposite direction till they reach their respective cellular halves; depolymerization of ParM filament then completes the process. During this depolymerization, ParM–ADP is liberated, followed by rejuvenation to ParM–ATP for another round of polymerization. The filament, thus grows till the target replicon is deposited in the two daughter cell halves (Fig. 6.12).

ParM belongs to a superfamily of ATPases that includes actin and its putative bacterial ancestor, MreB. MreB has been implicated in determining the bacterial cell shape. ParM, therefore, functions as a mechanochemical enzyme that couples ATP hydrolysis with active movement of plasmid DNA after replication. This as we discussed earlier is brought about by dynamic polymerization (filament formation) and depolymerization (filament break down) of ParM.

Interestingly, bacteria seem to have evolved different segregation mechanisms to faithfully partition their genetic components into daughter cells during division. While we have discussed above the role of actin-like homolog, in some Gram-positive bacteria, unique *par* locus was identified. Several plasmids in *Bacillus thuringiensis*, such as pBtoxis, and in *B. anthracis*, such as pXO1 seem to carry tubulin-like protein for this purpose (see Fig. 6.10). The *par* locus on such plasmids encodes TubZ, a cytoskeletal protein harboring tubulin signature motif, with high GTPase activity, and a small DNA-binding protein, TubR. TubZ was initially discovered as a protein involved in replication of the plasmid, pXO1, and was called as RepX. RepX has been shown to undergo GTP-dependent polymerization into long filaments, with GTP hydrolysis implicated in depolymerization of the fibers, and exhibit a GTP-dependent non-specific DNA binding. It thus appears to provide the simplistic spindle based on tubulin, and constitutes yet another mechanism (type III) of plasmid partitioning.

6.5.3 Maintenance of Monomers

Many plasmids may also have to invoke mechanisms to ensure that they remain in monomeric form. If a plasmid exists as multimer, it may suffer an improper segregation. The effect of such an improper distribution can be absolutely suicidal, especially for low copy number plasmids, as at every cell division one cell may receive none (cured), and the other all the plasmids. The classical example consists of ColE1 derivatives that form multimers in *recA*$^+$ *E. coli*. Many other plasmids, such as ColE1, P1, RK2, etc., maintain themselves as monomers and thus

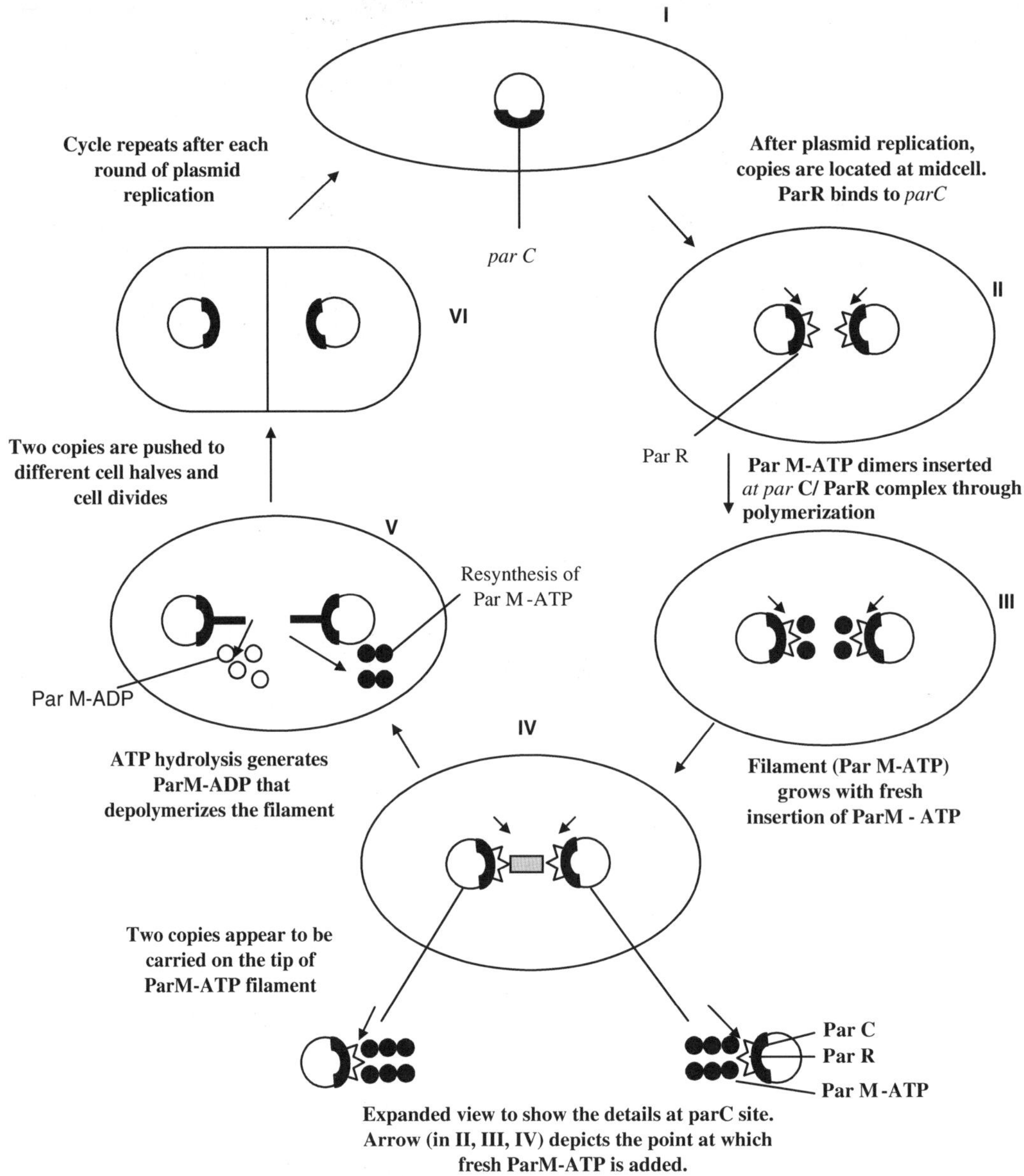

Fig. 6.12 Partitioning of plasmid R1 (Type II) post-replication

are not lost. The possible mechanism is the operation of a site-specific recombination that resolves the multimers.

In ColE1 plasmid, there is a cis-acting site *cer* where *E. coli* encoded Xer recombinase acts. The multimers of ColE1 have multiple direct repeats of *cer* and these are involved in Xer-mediated recombination that resolves them into monomeric units. Multimer formation takes place in $RecA^+$ strain but not in $RecA^-$, suggesting the involvement of homologous recombination. The *cer* is located in a 280 bp region and site-specific recombination takes place within a 35 bp stretch lying just outside the *cer*. Many ColE1-related plasmids such as, pMB1, ColK, and CloDF13 also employ a similar mechanism.

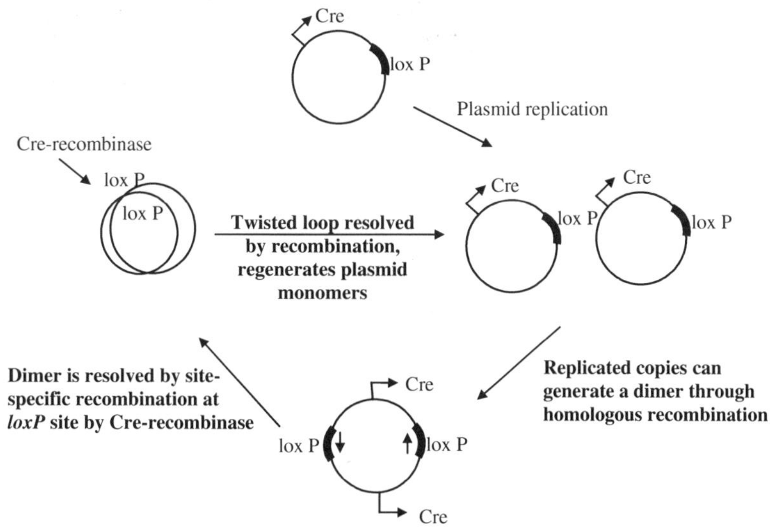

Fig. 6.13 *Cre-lox* site-specific recombination system of phage P1 to generate plasmid monomers from a dimer which are segregated at the time of cell division. Many plasmids use a similar mechanism but their sites and the recombinase and additional proteins required differ from plasmid to plasmid

Broad-host range plasmids, RK2, and RP4 carry *parCBA* and *par* function, respectively. They code for a resolvase to generate plasmid monomers from multimers. The plasmid pT181 carry two sites, RSA and RSB. RSA is used as a site for recombination between different plasmids as also within the co-integrates. Many pT181-like plasmids, such as pE194 and pSN2 family also share similar mechanism.

A very distinctive site-specific recombination (*cre-lox*) system operates in P1. It consists of two components: a *loxP* site at which crossing over takes place, and a phage encoded *cre* recombinase. No other phage or host function is required for this site-specific recombination. In rare instances, *cre*-mediated recombination may take place between *loxP*, and *loxB* site located on *E. coli* chromosome, leading to phage integration in the host genome. The *loxP* consists of two perfect 14-bp inverted repeats, separated by an 6-bp spacer, where crossing over takes place. Thus, when plasmid replicates, homologous DNA copies can recombine to give a dimer. Such a dimer will carry two *loxP* sites in direct orientation that will be engaged by the *cre*-recombinase and be resolved in unit copy replicons, prior to cell division (Fig. 6.13).

6.5.4 Post-segregational Killing/ Plasmid Addiction Systems

Many plasmids have evolved elaborate mechanisms to resist curing, thus ensuring their maintenance. In this, the rate of the plasmid loss is not affected but plasmid-free cells do not survive—a phenomenon referred to as post-segregational killing or plasmid addiction system. Though mechanisms may differ in final details, they are principally based on a toxin-antitoxin/ antidote production by the plasmid bearing cells. While the toxin is highly stable, the antidote is short-lived and needs to be continually synthesized. A cell that loses a plasmid may contain active toxin much longer, but the antitoxin disappears fast leading to the killing of the cured cell (Fig. 6.14). Though many plasmids following such a pathway have been described by now, we shall discuss some of the examples that have been worked out in detail.

In IncFII plasmid, R1, this system is referred as *hok-sok-mok*. A plasmid gene called *hok* (host killing) codes for a 52-amino acid protein that causes rapid cell killing. This action is countered by another gene, *sok* (suppression of killing) that codes for an untranslated anti-sense RNA. For

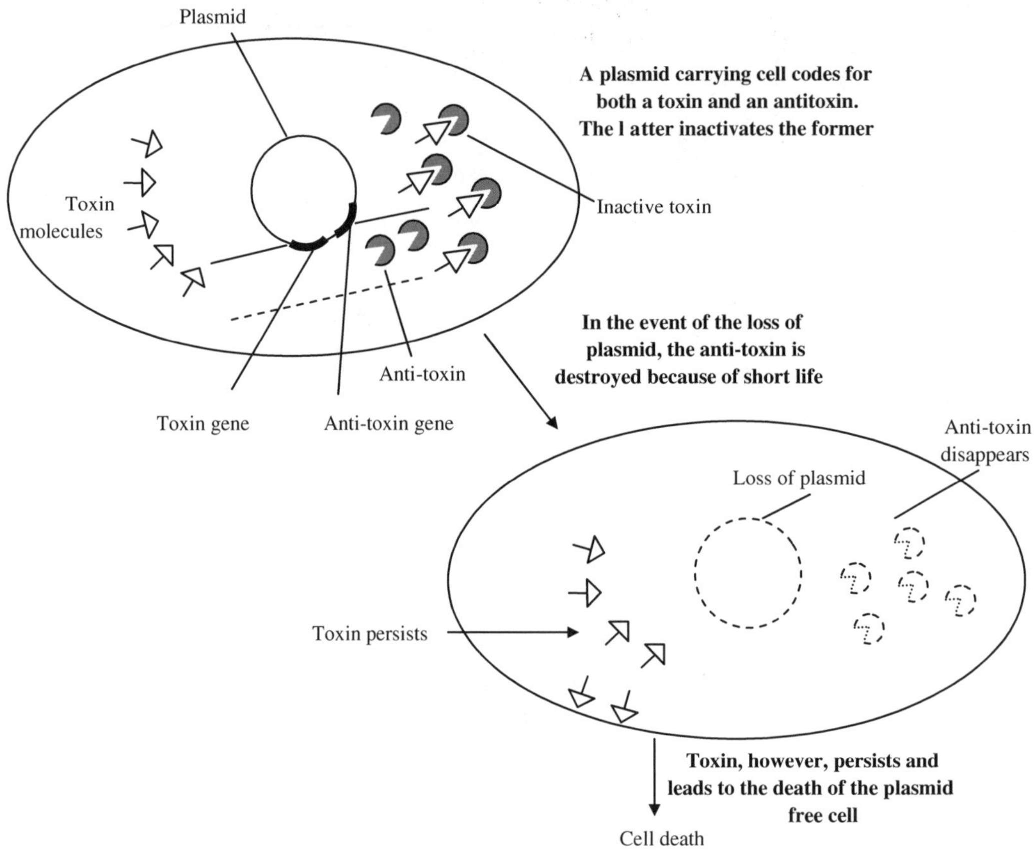

Fig. 6.14 Generalized mechanism of post-segregation killing of a cell when it loses a plasmid. A plasmid codes for both a toxin and an anti-toxin. In a plasmid containing cell, both are produced and the antitoxin inactivates the toxin. If the plasmid is lost, anti-toxin disappears fast because it is unstable. The toxin, however, persists and kills the plasmid-free cell

this function to be operative, at least 90 nucleotides of *sok* RNA are required. In a plasmid-carrying cell, therefore, *hok* mRNA is bound by *sok* anti-sense RNA making the former untranslatable and thus ineffective. In a plasmid-free cell (cured), while fresh synthesis of both *hok* and *sok* mRNA will be suspended, the *hok-sok* RNA hybrids will stay in the cytoplasm. The *sok* RNA, however, is short-lived and will be rapidly degraded, releasing, and leaving the *hok* mRNA to get translated and produce its lethal effect.

Another gene required for the full operation of this system is *mok* (mediator of killing). The *mok* is transcribed as part of the *hok* mRNA. The secondary structure of Mok-Hok mRNA is such that translation initiation site of Hok becomes available only when upstream Mok has been translated. Moreover, the 3′ end of the Mok-Hok mRNA has a structure that could fold back thus preventing ribosome's access to Shine–Dalgarno sequence and, therefore, blocking Mok translation. The *hok* expression is, therefore, brought about by slow processing of the 3′ end of *mok-hok* mRNA so that Mok and eventually Hok can be translated. Sok RNA has the ability to bind to all the three forms of Mok–Hok mRNA: full length RNA, RNA truncated at the 3′ end to facilitate Mok translation, and RNA whose secondary structure has already been altered by Mok translation, with the highest preference shown for the last form (Fig. 6.15). Hok protein's lethal action is brought about by its ability to disrupt the proton gradient across the membrane, inhibiting oxygen uptake.

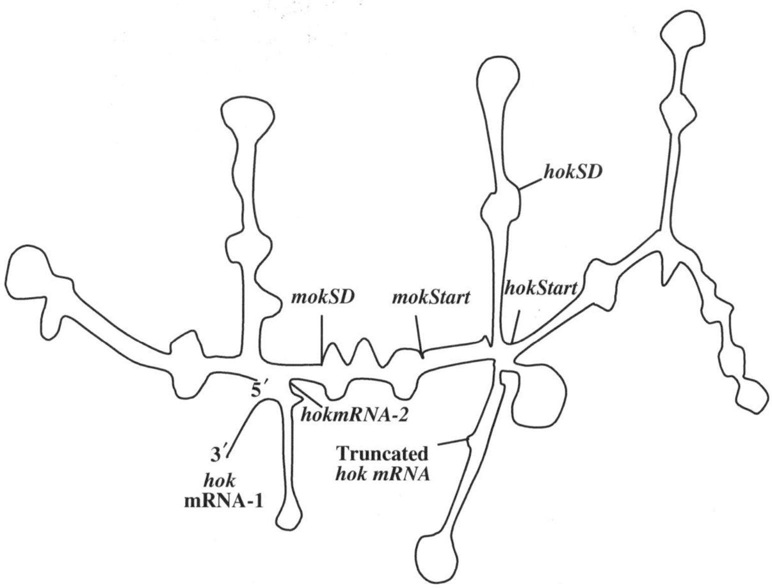

Fig. 6.15 Diagrammatic representation of the secondary structure of *mok-hok* mRNA of IncFII plasmid. The *sok* RNA binds to all the three forms: *mok-hok* full-length mRNA, RNA truncated at its 3′ end that allows Mok translation, and mRNA whose secondary structure has been altered by translation of Mok, with greatest affinity for the third form. The *sok-hok* RNA duplex does not allow Hok translation and is a target for RNase II digestion

A similar toxin–antitoxin system also operates in F Plasmid. It was known that loss of F plasmid may lead to filament formation and ultimately cell death. F plasmid carries a *ccd* (control of cell death) operon located near *oriS*. This operon consists of two genes, *ccdA* and *ccdB*, coding for two proteins, CcdA and CcdB. Of these two, CcdB is a long-lived stable toxin and CcdA is the unstable antitoxin, requiring continuous production for its functioning. The *ccd* operon is autoregulated by CcdB and CcdA stimulates the binding of CcdB to its promoter. In the plasmid-carrying cells, the CcdB acts by complexing GyrA protein and CcdA overcomes this action by releasing GyrA from such a complex. In a plasmid-free cell, however, the unstable CcdA quickly disappears leaving the toxin to carry out its lethal action.

In *Thiobacillus ferroxidans* broad-host range plasmid, pTF-FC2, three components are involved: *pasA* coding for the antidote, *pasB* for bactericidal toxin, and *pasC* that encodes a protein that enhances the neutralizing effects of antitoxin. Many other plasmid maintenance systems based on post-segregational killing have now been described from a variety of plasmids.

6.5.5 Restriction-Modification System

The evolution of restriction-modification system has long been thought to be a mechanism by which cells protect themselves from foreign DNA, such as plasmids and the phage. This involves a lethal restriction endonuclease having the ability to cleave double-stranded, non-self DNA. This lethal action is not exercised on resident (self) DNA because of the action of methyltransferase that modifies the DNA. Hundreds of restriction enzymes have been discovered, their genes cloned, and sequenced. One interesting observation has been that many of these genes are located on naturally occurring plasmids, including phage P1.

This raises a very pertinent question. Could this have a role in plasmid maintenance? The question becomes more important, as methyltransferases are less stable than restriction endonucleases. Thus, if a cell does not receive a copy of the plasmid, endonuclease may remain active much longer to bring about its toxic effect. However, their location on the plasmid may be for more of a mutualistic reason. While on one hand, location on the plasmid may improve their expression (based on the copy number), providing protection from the foreign DNA, on the other, it may provide a mechanism to ensure plasmid maintenance.

6.6 Host Range

As the copy number is regulated by the plasmid, the types of plasmids, existing in a bacterial system are also a plasmid-specific property. Based on their distribution pattern, plasmids have been classified as broad host range or narrow host range plasmids.

The broad host range plasmids such as RK2 and RSF1010 may employ different strategies to survive in many bacterial host species. These strategies are generally related to their replicational requirements. For example, plasmid RK2 uses *oriV* and a plasmid encoded protein TrfA to replicate in *E. coli*. A mutation in *oriV* may abolish its replication and, therefore, survival in *E. coli*. But this mutation can help the plasmid to establish itself in *Pseudomonas aeruginosa,* as there it may utilize a different origin of replication. Although RK2 codes for very few replication-required functions, it seems to have evolved a way to utilize different proteins from different host species. Plasmid RSF1010, on the other hand, codes for majority of the proteins needed for its replication. This functional autonomy permits the plasmid to thrive in different host backgrounds.

6.7 Plasmid Incompatibility

A bacterial cell can carry different types of plasmids and transmit each plasmid faithfully to the daughter cells at the time of cell division. However, this property is not universally applicable and some plasmids even if transferred under a strong selection cannot be maintained stably thereafter. Thus, the plasmids that can coexist stably are referred to as compatible and those that cannot are called incompatible plasmids.

The most thoroughly studied basis of incompatibility is the sharing of replicational and partitioning requirements by two types of the plasmids. In such a case, a competition may be exercised on the utilization of the initiator protein that may become limiting. Thus, if a cell contains two types of plasmids, say type I and type II, at the time of replication, one type will be picked up randomly. If type I is replicated, for example, there will be two copies of type I and only one of type II plasmid. In the next round, type I will have higher probability of being replicated and plasmid type II may gradually be lost. A similar outcome is expected if the two plasmid types share the same pathway for partitioning.

Incompatibility is such a universal property of the plasmid that it has turned out to be most suitable for plasmid classification. Plasmid incompatibility was first described for the F plasmid in *E. coli* strains in the early 1960s. The existence of this phenomenon was indicated by the observation that male *E. coli* cells can contain the F plasmid either in the free, autonomous, or integrated state, but the two situations are never found in the same cell. The same is also applicable to two free F plasmids. The inference drawn from this observation was subsequently validated experimentally.

In the early 1970s, Datta and Hedges introduced a formal system of classification based on the incompatibility response of the plasmids under question. Since then this has been further expanded and reviewed by several workers. Readers interested in the details of this aspect can consult many such reviews, such as that by Datta (1975, 1979), Novick (1987), and Couturier et al. (1988). The basis of this classification, is that plasmids incompatible with each other are assigned to the same incompatibility group (*inc*). So far $\sim$30 *inc* or incompatibility groups have been recognized among enteric bacteria, and $\sim$7 such *inc* groups have been reported among staphylococcal plasmids. To name a few, they are: IncL/M, IncN, IncP, IncT, IncU, IncW, IncY, IncB/O, IncFI, FII, FIV, FV, FVI, IncI1, I2, Ir, IncK, Com9, IncH11, H12, H13, IncX, IncA/C, IncD, IncJ, IncV, etc. Thus, any new plasmid identified and described is classified on the basis of the incompatibility relationship with the known plasmids, and this may even lead to the creation of an entirely new class of incompatibility group.

The plasmid incompatibility is an important consideration in designing any plasmid transfer experiment as well as in selecting a cloning vector for genetic engineering, as both demand the stability of the introduced plasmid.

6.8 Plasmid Amplification

As discussed earlier, many plasmids require de novo protein synthesis for initiation of every fresh round of DNA replication. The ColE1 plasmids are virtual exception to this rule. As ColE1 plasmids do not require new protein synthesis for replication, their copy number can be preferentially amplified relative to the copy number of the chromosome. For example, if cells containing a ColE1 plasmid are treated with chloramphenicol, a protein synthesis inhibitor, chromosomal replication will stop, but the plasmid replication will continue. This leads to enhanced number of plasmid copies. Many popularly used cloning vectors, such as pBR322 and pUC, and several others contain ColE1 origin of replication.

6.9 Plasmid Transfer

One distinctive property of the plasmids that led to their widespread usage in genetic analysis, gene cloning as well as in microbial evolution, is based on their ability to get transferred between bacteria, sometimes even promiscuously.

That a gene/(s) is/are located on the plasmid would often be demonstrated by their transfer to a cell that does not possess the same plasmid. If selection is made for more than one marker, an en bloc transfer would be a characteristic property of the plasmid-based transfer. The effectiveness of the plasmid transfer, in fact, led to the discovery of conjugation in bacteria, such as in *E. coli*.

The plasmid transfer during bacterial mating requires four distinct steps:

1. Cell-to-cell contact between donor and recipient cells (effective contact)
2. Preparation of donor DNA transfer (mobilization)
3. Transfer *per se*, and
4. Replication and restoration of the plasmid both in the donor and the recipient.

Plasmids can be classified into different categories based on these requirements. A non-conjugative\non-transmissible plasmid cannot generate effective contact. A conjugative plasmid can successfully bring about effective contact. A transmissible plasmid can prepare its DNA for transfer, and a self-transmissible plasmid is both conjugative and transmissible. As conjugative functions are often not plasmid-specific, one plasmid can help the other plasmid in its transfer. For example, F plasmid is transmissible and ColE1 plasmid is mobilizable but non-conjugative. If a cell carries both F and ColE1 plasmids, ColE1 can be transferred making use of the contacts generated by F, a process known as donation. In contrast, mobilization function is plasmid-specific, such that a self-transmissible plasmid cannot effect the transfer of a non-mobilizable plasmid by a simple genetic complementation. The transfer of the latter, however, can be helped if the two plasmids recombine, a process referred to as conduction. F-mediated transfer of *E. coli* chromosome can be cited as an example of conduction.

Plasmid transfer is a multigene function and the plasmids carry a whole *tra/trb* operon to provide all the inputs. The *tra/trb* operon often occupies a major chunk of the plasmid genome. The classical model system of a transmissible plasmid and the intercellular plasmid transfer is the fertility plasmid, F of *E. coli*, which we have discussed earlier in the chapter on Conjugation.

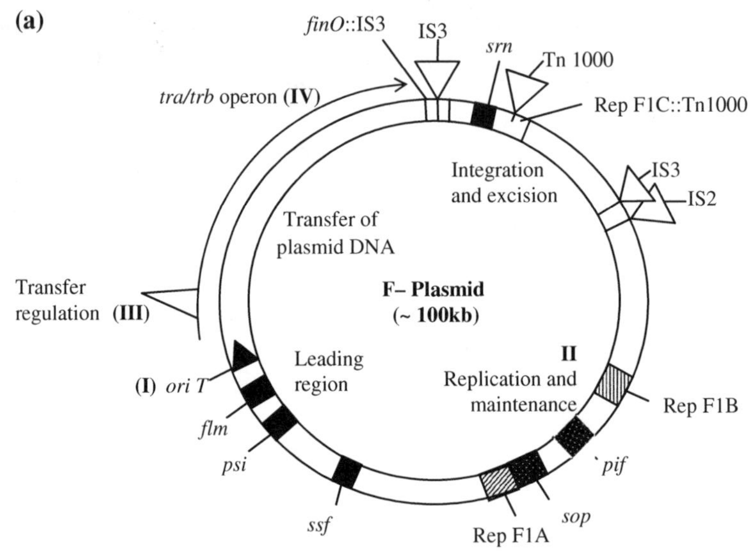

Fig. 6.16a Plasmid F—the prototype model. Generalized map of F plasmid. Specific functional region demarcation has been depicted. The plasmid carries three *ori* regions. RepF1A, RepF1B, and RepF1C. Of these, RepF1A is the predominant *ori* consisting of both bidirectional (*oriV*) and unidirectional (*oriS*) origin of replication. RepF1B is a secondary *ori* and is used only in the absence of RepF1A. The third, RepF1C is inactive due to insertion of Tn1000; *srn*: inducible modifier of host membrane; *pif*: phage inhibition; *sop*: plasmid partitioning system; *ssf*: single-strand binding protein; *psi*: transiently expressed following its transfer, prevents SOS response by not allowing activation of Rec; *flm*: post-segregation killing; *ori T*: origin of transfer; *tra-trb* operon genes involved in various steps of plasmid transfer

6.10 F Plasmid: A Prototype Model System

As described earlier, F, a classical model of plasmid function and transfer, is a large plasmid of ~ 100 kbp in size. The detailed map of the plasmid shows a typical functional organization, consisting of four different regions (Fig. 6.16a). Beginning clockwise, the first ~ 15 kb region consists of four copies of transposable elements: two copies of IS3, one copy of IS2, and one copy of Tn1000. While one copy of IS3 is integrated in *finO* gene, Tn1000 is inserted in one of the replication locus (RepF1C), both events leading to inactivation of the respective genes. This region is involved in all integration-excision events. The *srn* operon situated close to RepF1C codes for one of the toxin–antitoxin system on F (Fig. 6.16a). Following this is a region from which no gene has yet been described. The next region is devoted toward plasmid replication and maintenance, and thus involves multiple functions (Fig. 6.16bII). It consists of two replication loci, RepF1A and RepFIB. While the latter is partly defective, RepF1A constitutes the primary replication system. RepF1A is made up of both bidirectional (*ori1*), also called *oriV,* and unidirectional *ori* (*ori2*), referred to as *oriS.* Also present in this region is *repE,* whose product, RepE is required for plasmid replication, along with the copy number control region, *incC.* Another interesting feature is the presence of a partitioning system consisting of *sop* operon, made of *sopA* and *sopB*, whose products act at *sopC* to faithfully distribute the plasmid copies to daughter cells at the time of cell division, as described earlier. True to its low copy number status, F also codes for two additional mechanisms of plasmid maintenance, one consists of *resD,* coding for a resolvase that acts at *rsfF* site, as part of the site-specific recombination system to maintain plasmid monomers. The other consists of toxin–antitoxin system, *ccdA–ccdB* that leads to post segregational killing of cells that does not

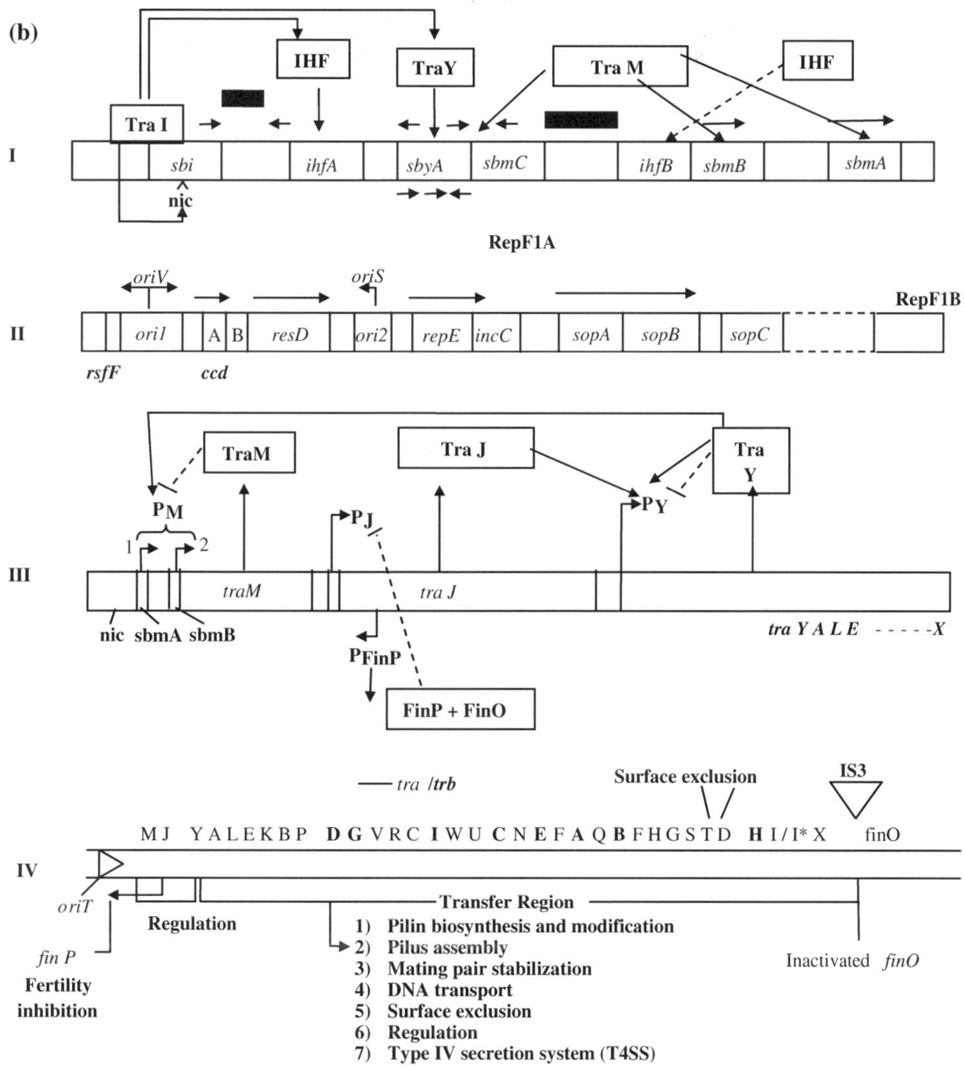

Fig. 6.16b Area marked *I-IV* in Fig. 6.16a are expanded to provide some details. I—*ori T* region containing direct and inverted repeats (marked by *arrows*) and intrinsic bend (▬) regions, bound by TraI, TraY, and TraM along with the integration host factor (IHF) are also depicted. They generate a relaxosome. The *nic* represents the site where single-strand nicking takes place with the help of TraI and supported by TraY. II—Replication and plasmid maintenance region—contains the pre-dominant *ori*F1A which comprises both bidirectional and unidirectional replication *ori*, and the *ccd* and *sop* operons required for plasmid maintenance. A site-specific recombination system (*rsfF*), also required for maintenance has been identified. A secondary replication origin (RepF1B) is present but is generally not utilized. III—transfer regulation system—principally involves TraM, TraJ, and TraY. Transcription of *tra/trb* operon from PTraY is positively regulated by TraJ. Both TraM and TraY are autoregulated and TraY also activates P_M. TraJ is repressed by anti-sense FinP RNA which in F-like plasmids is further facilitated by FinO. IV—The structure of *tra-trb* operon—the operon codes for a large transcript from P_Y covering about 33.3 kb region and providing multiple functions shown. Several other promoters have also been identified in this region. The *finO* is the last gene which is inactivated in F due to insertion of an IS3. *trb* genes are shown in bold. The preceding region is oriT with regulatory genes (see part III)

receive a copy of F. We have discussed all these mechanisms earlier in this chapter. You may recall that F^+ cells do not allow the "female-specific" phages to grow on them, and the locus responsible for this is *pif,* consisting of *pifC* and *pifA*.

The following ~13 kbp region that culminates in *oriT* is referred to as leading region, as this is the first region to enter the recipient cell, when transfer begins at *oriT*. Very few genes have been identified in this region and their roles have been implicated in conjugation. For example, the *psiB* gene product is transiently expressed in the recipient, upon its transfer, and is envisaged to inhibit SOS response by preventing the activation of RecA co-protease activity. The *ssf* gene is similar to the host *ssb*, encoding a single-strand DNA binding protein that might protect the single-stranded DNA entering the recipient cell prior to its replication. The last component of this region marks the *oriT*. F *oriT*, ~350 bp in size is complex, being made of structural elements such as direct repeats (DR), inverted repeats (IR), intrinsic bends, and A- tracts. In addition, it contains the *nic* site, and binding sites for TraI (*sbi*),TraY (*sbyA*),TraM (*sbmC, B, A*) and IHF(*ihfA* and *ihfB*) (Fig. 6.16bI). As shown in this figure, the *nic* site is located on the bottom strand and all protein binding sites and most of the structural elements are located 5′ to *nic*, The final segment of the F plasmid is the large ~33 kbp transfer region, occupied by the *tra/trb* operon. The constituent genes of this operon are involved in different steps of conjugational DNA transfer (Fig. 6.16bIV) that we have discussed in Chap. 3 on Conjugation. The last gene is *finO*, which in F is inactive due to insertion of IS3.

During conjugation, a single DNA strand is transferred from a donor cell to a recipient cell. This multistep process is mediated by the genes of the transfer region. The first activity is initiated at *oriT*, where a protein complex called relaxosome is organized. The constituent proteins of relaxosome, TraY and IHF bind to *sbyA* and *ihfA* sites, respectively, and bend DNA. Such a conformational change may facilitate the creation of a single-stranded region at *nic* to which TraI could bind. These auxiliary proteins thus recruit TraI to the *nic* site. TraI is a relaxase, nickase, and a helicase, and nicks the DNA to generate the single-stranded DNA end for further reaction. TraM which also has multiple (three) binding sites in this region is not necessary for nicking but perhaps facilitates the generation of the relaxosome and prepare DNA for the transfer (Fig. 6.16bI). While TraI causes the nicking it gets covalently attached to the 5′end of the DNA and remain so throughout the transfer process. At the same time, its helicase activity also unwinds ~200 bp DNA stretch. The *traM* gene is immediately adjacent to *oriT*, and its product TraM, an auto-regulated protein, is essential for DNA processing during transfer (Fig. 6.16bIII). It regulates F DNA transfer by performing a crucial signaling function that may involve sensing the stable mating pair formation. TraM is also shown to interact with TraD, a reaction crucial for F conjugation. TraM forms a link between the events initiated at *oriT*, and a key component of the conjugative pore, the coupling protein, TraD, that facilitates the proper positioning of the plasmid at the conjugative pore. TraD, therefore enables the relaxosome to be linked to another protein complex, called transferosome or type 4 secretion system (T4SS). In other words, the TraI bound DNA is transferred to T4SS by TraD. Several genes of *tra/trb* operon code for the constituents of T4SS. Thus, upon mating pair formation, generation of a nicked DNA, and the covalent attachment of TraI to its 5′ end and proper positioning of the plasmid, single-stranded DNA with attached TraI is transported into the recipient cell. The transfer then continues in a time-dependent manner. Several TraI molecules are involved not only in initiation but also in completion of the transfer process. In the recipient, TraI is translocated from its 5′ location on single-stranded DNA to 3′ end, which allows not only the complementary strand synthesis but also leads to joining of the two ends. The joining reaction involves an oligonucleotide with a hairpin, a structure that has been detected in F *oriT*. In the donor on the other hand, the 3′end of the cleaved plasmid serves as the primer for PolIII which initiates complementary strand synthesis.

Successful transfer requires multiple functions that include pilin biosynthesis and modification, pilus biogenesis, signal transduction, mating pair formation and stabilization, DNA processing, and its transfer (Fig. 6.16bIV).

The whole *tra/trb* operon is regulated by three main regulators, TraM, TraJ, and TraY forming a regulatory loop and the respective genes are located in the beginning of the operon (Fig. 6.16bIII). While *traM* and *traJ* are the two mono-cistronic genes, *traY* constitutes the first gene of *traY–traX* stretch. Both *traM* and *traY* are autoregulated. The mechanism through which these regulators function, is referred to as "latch relay" mechanism, wherein TraJ activates *Py* that gives rise to TraY. TraY, has a positive effect on P_m, and transcripts from P_m read-through into *traJ*. Transcription of *tra/trb* operon from P_y is positively regulated by TraJ, thus stimulating the transfer process. (Fig. 6.16bIII). The *traJ* is subjected to another level of regulation that is referred to as fertility inhibition and is described below. If one presumes that all the regulatory check points for F transfer are provided by the plasmid itself, this will amount to oversimplification. F and other narrow host-range plasmids have a superimposing regulation involving a panoply of host-encoded factors. These factors can act at the DNA level (such as activators, repressors, architectural proteins, excision and recircularization effectors, and Dam methylation); the RNA level (RNase susceptibility of antisense RNA and its target); and protein level (inhibitor binding, proteolysis, and conformation change).

6.11 Other Properties

Many plasmids besides carrying genes for essential functions also exhibit other properties.

6.11.1 Surface Exclusion

Many plasmids limit the ability of the cells containing them to act as recipients, a phenomenon referred to as surface exclusion. F plasmid carrying cells, for example, can act only as donor but not as recipient of the same. To operate surface exclusion, F plasmid carries two genes, *traS* and *traT*. The *traS* encodes a 16.9 kDa inner membrane protein, which when present in a cell (F-containing) appears to prevent the triggering of the conjugal DNA metabolism in the donors. In other words, *traS* blocks the transmission of a mating signal between two F-containing cells. The *traT*, on the other hand, codes for a 26 kDa lipoprotein that constitutes a major outer membrane protein component of F-containing cell. Present in ~20,000–30,000 copies per cell, it appears to function by inhibiting the mating aggregate formation.

6.11.2 Fertility Inhibition

As discussed earlier, F plasmid transfer and its spread from F^+ to F^- cells occur rapidly, something that could be referred akin to infectious transfer. In comparison, the transfer of R plasmid occurs at a frequency of about 0.02 % of a population of cells. The kinetics of transfer shows that an R^- cell that may have just received the R plasmid is competent to transfer the same immediately. At a later time, however, i.e., from the cell containing the plasmid for a long time, the transfer is inhibited due to what is known as fertility inhibition. This is based on the synthesis and action of a multisubunit repressor coded by the *fin* genes and its repressive action on transcription of the genes required for transfer.

In cells harboring F or F-like plasmids (such as R), *tra* gene expression and hence conjugative transfer may be repressed by the two *fin* genes, *finP* and *finO*. The *finP* codes for a small antisense RNA, and the *finO* (most distal gene of *tra* operon)-encoded polypeptide that leads to the inhibition of the expression of *traJ*. TraJ being the positive regulator of P_{traY} and, therefore, of *tra/trb* operon, in turn, precludes transcription from P_{traY}. The FinP antisense RNA is ~78 nts in length and is transcribed from P_{finP} constitutively, located within and in opposite orientation to the un-translated leader region of *traJ*.

Complementary pairing between *finP* and *traJ* mRNAs, generates an RNA–RNA duplex, which is cleaved by RNase III, and prevents the translation of *traJ* mRNA. This leads to non-expression of the rest of the Tra functions. FinO seems to exert its co-regulatory effect on *traJ* expression by stabilizing FinP anti-sense RNA. FinO product has been shown to be an RNA-binding protein. F plasmid does not exhibit fertility inhibition due to the inactivation of *finO* by insertion of an IS3, such that F transfer remains derepressed.

6.11.3 Plasmids in Gram-Positive Bacteria

Plasmids are also common in Gram-positive bacteria as they are to Gram-negative members. Some members such as, *Enterococcus faecalis* and *Streptococcus lactis* may carry several plasmids whereas many others may not carry any. In *E. faecalis*, conjugative plasmids bearing fertility factors and novel transposons have been reported. You may recall that linear plasmids have also been described in some Gram^{+} bacteria. Besides these properties, some plasmids may also confer resistance against those bacteria not possessing them. One interesting type of the plasmid described from *E. faecalis* confers a mating response to peptide sex pheromones excreted by the plasmid containing cell.

In *Enterococcus faecalis* (*E. hirae*), conjugative plasmids have been found. They have been grouped in two categories: one (e.g., pAD1, pOB1, pPD1, pJH2, pAM1, pAM2, and p AM3) get transferred at a relatively high frequency in broth (10^{-3}–10^{-1}/donor), and the other (such as pAC1, pIIP501, and p5M15346) are transferred poorly in broth but efficiently in filter matings. This difference involves the production of sex pheromones by streptococci. Such pheromones, produced by the so-called recipient cells are soluble small peptides (7–8 amino acids) that induce certain plasmid carrying donor cells to adhere and facilitate the formation of donor-recipient mating aggregates. The role of the plasmid is to respond to a pheromone produced and released by the recipient in a very specific manner. Thus, stimulation of only some type of donor for conjugation will take place in a population, which in turn is determined by the type of the plasmid. For this response to materialize, the chromosomally coded and recipient released pheromone diffuses into the donor cell, where it reacts with a plasmid-coded product, called responding substance. This in turn, induces the production of aggregation substance (AS), the gene for which is also located on the plasmid. The latter, in fact, facilitates the formation of donor recipient aggregates, essential for the plasmid transfer. The plasmid also carries a specific inhibitor gene that represses the chromosomal pheromone gene in the donor, thus not allowing the endogenous production of the pheromone (see Fig. 3.19). Once the plasmid gets transferred in the recipient, the pheromone-based circuit is broken and the plasmid located inhibitor gene will suppress any further pheromone synthesis. We have discussed this process under conjugation as well.

Further Readings

Actis LA, Tolmasky ME, Crosa JH (1998) Bacterial plasmids: replication of extra-chromosomal genetic elements encoding resistance to anti-microbial compounds. Front Biosci 3:43–62

Bouma JE, Lenski RE (1988) Evolution of a bacterial/plasmid association. Nature 335:351–352

Chattoraj DK (2000) Control of plasmid DNA replication by iterons: no longer paradoxical. Mol Microbiol 37(3):467–476

de la Cueva-Mendez G, Pimental B (2007) Genes and cell survival: lessons from prokaryotic plasmid R1. EMBO Rep 8(5):459–464

del Solar G, Giraldo R, Ruiz-Echevarria MJ, Espinosa M, Diaz-Orejas R (1998) Replication and control of circular bacterial plasmids. Microbiol Mol Biol Rev 62(2):436–464

De-Nap JCB, Hergenrother PJ (2005) Bacterial death come full circle: targeting plasmid replication in drug-resistant bacteria. Org Biomol Chem 3:959–966

Dionisio F, Matic I, Radman M, Rodrigues OR, Taddei F (2002) Plasmids spread very fast in heterogeneous bacterial communities. Genetics 162:1525–1532

Eckhardt T (1978) A rapid method for the identification of plasmid deoxyribonucleic acid in bacteria. Plasmid 1(4):584–588

Gerdes K, Howard M, Szardenings F (2010) Pushing and pulling in prokaryotic DNA segregation. Cell 141:927–941

Hinnebusch J, Tilly K (1993) Linear plasmids and chromosomes in bacteria. Mol Microbiol 10(5): 917–922

Hughes VN, Datta N (1983) Conjugative plasmids in bacteria of the 'pre-antibiotic' era. Nature 302:725–726

Larsen RA, Cusumano C, Fujioka A, Lim-Fong G, Patterson P, Pogliano J (2007) Treadmilling of a prokaryotic tubulin-like protein, TubZ, required for plasmid stability in *Bacillus thuringiensis*. Genes Dev 21:1340–1352

Novick RP, Clowes RC, Cohen SN, Curtiss R III, Datta N, Falkow S (1976) Uniform nomenclature for bacterial plasmids. Bact Rev 40(1):168–189

Novick RP, Hoppensteadt FC (1978) On plasmid incompatibility. Plasmid 1(4):421–434

Ringgard S, Ebersbach G, Borch J, Gerdes K (2007) Regulatory cross-talk in the double par locus of plasmid, pB171. J Biol Chem 262(5):3134–3145

Sengupta M, Austin S (2011) Prevalence and significance of plasmid maintenance functions in the virulence plasmids of pathogenic bacteria. Infect Immun 79(7):2502–2509

Svara F, Rankin DJ (2011) The evolution of plasmid-carried antibiotic resistance. BMC Evol Biol 11:130–139

Velappan N, Sbattero D, Chasteen L, Pavlik P, Bradburry ARM (2007) Plasmid incompatibility: more compatible than previously thought. Protein Eng Des Sel 20(7):309–313

Zatyka M, Thomas CM (2006) Control of genes for conjugative transfer of plasmids and other mobile elements. FEMS Microbiol Rev 21(4):291–319

7 Transposable Elements

7.1 General Considerations

DNA as a master molecule in the cell has generally been regarded as static, with the only movement associated being at the time of chromatid/chromosome separation during cell division. This notion stayed for a long time, and, in fact, allowed geneticists to generate extensive linkage maps wherein a gene was assigned a fixed location on the chromosome. Such maps have served many useful purposes in genetic analyses in all organisms in the years to come, and still constitute an important exercise in genetics. However, in the 1940s Barbara McClintock described genetic elements in maize, which had the capacity to move around the genome. Calling them "controlling elements", McClintock was able to show that their mobility greatly influences the gene expression.

Although a revolutionary idea, McClintock's discovery largely went unnoticed until 1960s, when three groups independently, described these mobile elements in bacteria. This discovery paved the way for detailed genetic analysis and the mechanism involved in this mobility, as the bacterial molecular genetics by this time had reached greater heights than their eukaryotic counterpart. These elements were collectively referred to as transposable elements (TEs) or mobile genetic elements. Before being understood fully, they acquired several epithets such as jumping genes, selfish DNA, and junk DNA. Many of these names stand retracted, as we understand these elements, their biology, and their role in biological evolution much better.

The important basis of their discovery in all the systems was the instability (mutations) caused by their movement. The rate of this instability was much higher than that could be explained by a normal point mutation. Moreover, the reversibility of such mutations remained largely unaffected by treatment with any known mutagen. With their discovery in bacteria they were subjected to detailed structure–function analysis, leading to a wealth of information. However, they appear to be ubiquitously present from bacteria to humans, suggesting the important role they have played in biological evolution as a whole.

Another striking information that came out of their discovery is that the geneticists/evolutionary biologists found an additional mechanism of generating genetic variability as these elements not only caused mutations but also sponsored chromosomal rearrangements. This has an important connotation as lower organisms majorly relied on creating variations in individual genes. The presence of these elements, therefore, would have hastened the process in the face of very low rates of gene mutation. Many of these roles are equally important for eukaryotic organisms as well. In this chapter, we deal with the bacterial TEs but readers interested in other systems can easily find the required material from various sources.

S. Srivastava, *Genetics of Bacteria*, DOI: 10.1007/978-81-322-1090-0_7, © Springer India 2013

In bacteria, these elements have been primarily classified into three categories: Class I elements or Insertion Sequences (IS), Class II elements or Transposons (Tn), and Transposable phages (such as Mu) that comprise class III elements. Transposons are further subdivided into type I, composite or type II, simple Tn of Tn3 family. All these elements have been responsible for generating gene mutations as and when they move into the reading frame of an active gene. In fact, many of the spontaneous mutations identified earlier in bacteria are ascribed to the mobility of these genetic elements. In addition, because of the chromosomal aberrations that accompany their movement, they are considered to be agents of in vivo genetic engineering.

All these TEs share several properties yet each of them is different and so are identified as distinct entities. Their similarities are:

(i) The TEs are constituent residents of bacterial genome; chromosome, plasmids, or even phages. From here on they have the ability to move from one DNA element to another or within the same DNA, from one site to another.
(ii) Their intercellular movement requires a mechanism of gene transfer (Fig. 7.1a,b).
(iii) They share a basic structural plan with the majority of them being flanked by inverted or direct repeat sequences (IR or DR) of short or fairly large size. Phage Mu is the only exception (Fig. 7.2).
(iv) They all code for functions, especially the enzyme transposase (Tpase) required for their movement. The process, therefore, is known as transposition.
(v) They may (Tn elements) or may not (IS elements) carry additional phenotypic markers that help in tracking them and their movement.
(vi) The ends of a transposable element are important because they serve as sites for Tpase action.
(vii) Their insertion into a target site is highly random with only a localized site selection specificity observed with some. One important exception is Tn7 which inserts into a specific target site, referred to as *attTn7*.
(viii) The insertion at a site is accompanied by target site duplication (TSD), the length and not the sequence of which is element-specific (Fig. 7.3). On the basis of this feature, these elements can be classified as those effecting 5, 9, 11, 13, or 14 bp duplications.
(ix) The transposition mechanism is either replicative, with both the donor and the new site carrying a copy each of the transposable element, or conservative (cut and paste), wherein the transposable element leaves the donor site and enters into a new site. This has often been used to categorize the TEs.
(x) These elements can also be excised from a site either precisely or imprecisely. While the former reverses the effect of insertion, the latter leaves a footprint at the site.
(xi) The mobility of these elements can generate mutations as well as chromosomal aberrations.
(xii) Their presence can also be ascertained by electron microscopy, as a result of intramolecular reannealing. This involves the denaturation of a DNA segment or a genetic unit carrying the TE, followed by rapid cooling to facilitate self-annealing. A stem-loop structure is formed, where the length of the stem indicates the IRs, and the loop is formed by the intervening region (Fig. 7.4).

7.2 Insertion Sequences (IS)

IS are the simplest of the bacterial TEs coding only for the function required for their transposition. Being made up of a specific length (ranging from ~800 bp to <2.5 kb), they are natural inhabitants of bacterial chromosome, their plasmids and phages (Table 7.1). When first described, they were few in numbers and thus based on their size, they could be classified as IS1, IS2, IS3, IS4, IS10, IS50, etc. Subsequently, however, there has been an explosion in their numbers and varieties that runs into several hundreds, a number which is likely to increase further due to genome sequencing information.

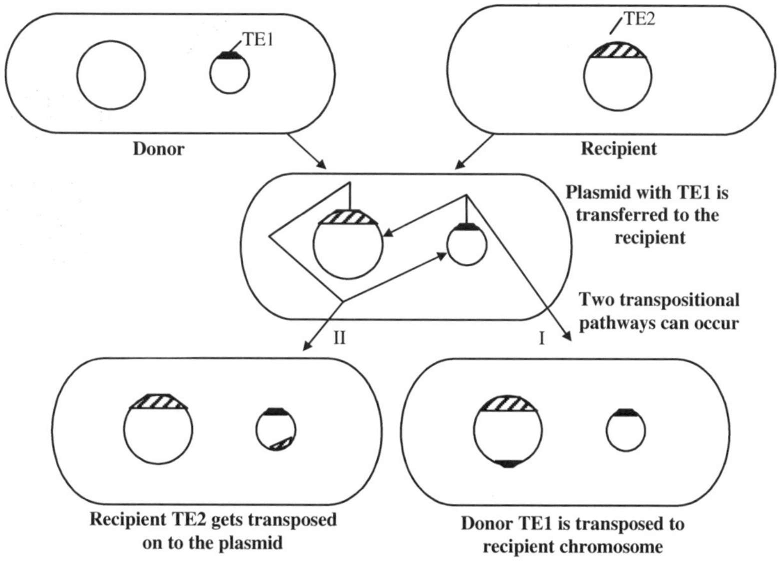

Fig. 7.1a Intercellular transfer of transposable element (TE) by conjugation: a donor cell carrying a transposon on a transmissible plasmid is mated to a recipient that may carry another transposon on the chromosome. The ex-conjugants so formed can be of two types: one in which plasmid-associated transposon moves on to the chromosome and chromosomal transposon gets transposed on the plasmid. Whether donor DNA (plasmid or chromosome) will retain the original transposon will depend upon the type of the transposition pathway (refer text). Also, it is not essential that the recipient chromosome will always carry a transposable element

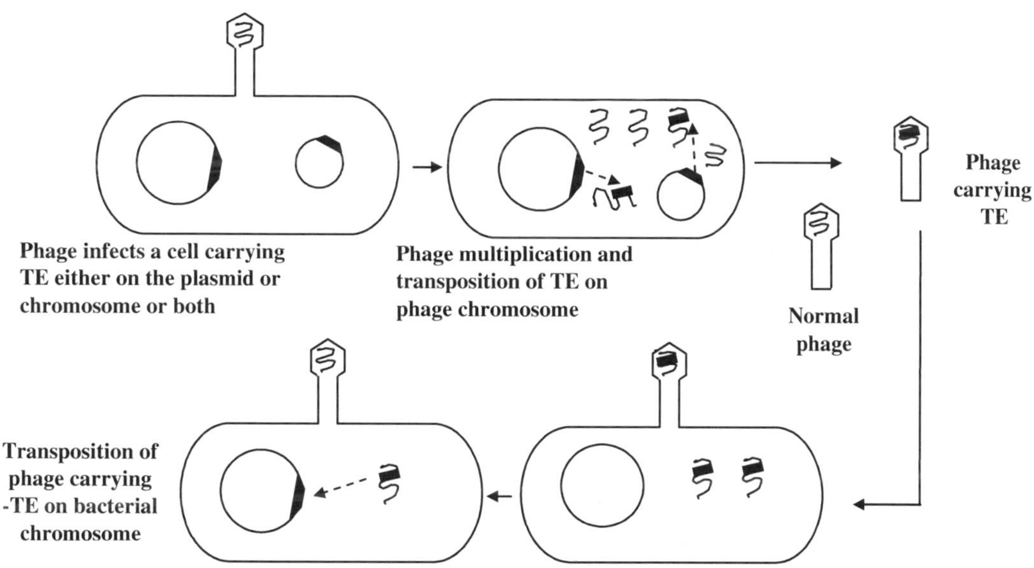

Fig. 7.1b Intercellular transfer of transposable element (TE) by a phage: a phage infects a cell that carries a TE on plasmid or on the chromosome. When phage multiplies, a copy of the TE may be transposed to the phage chromosome. When such a phage infects another cell this TE may move on to bacterial chromosome or a plasmid

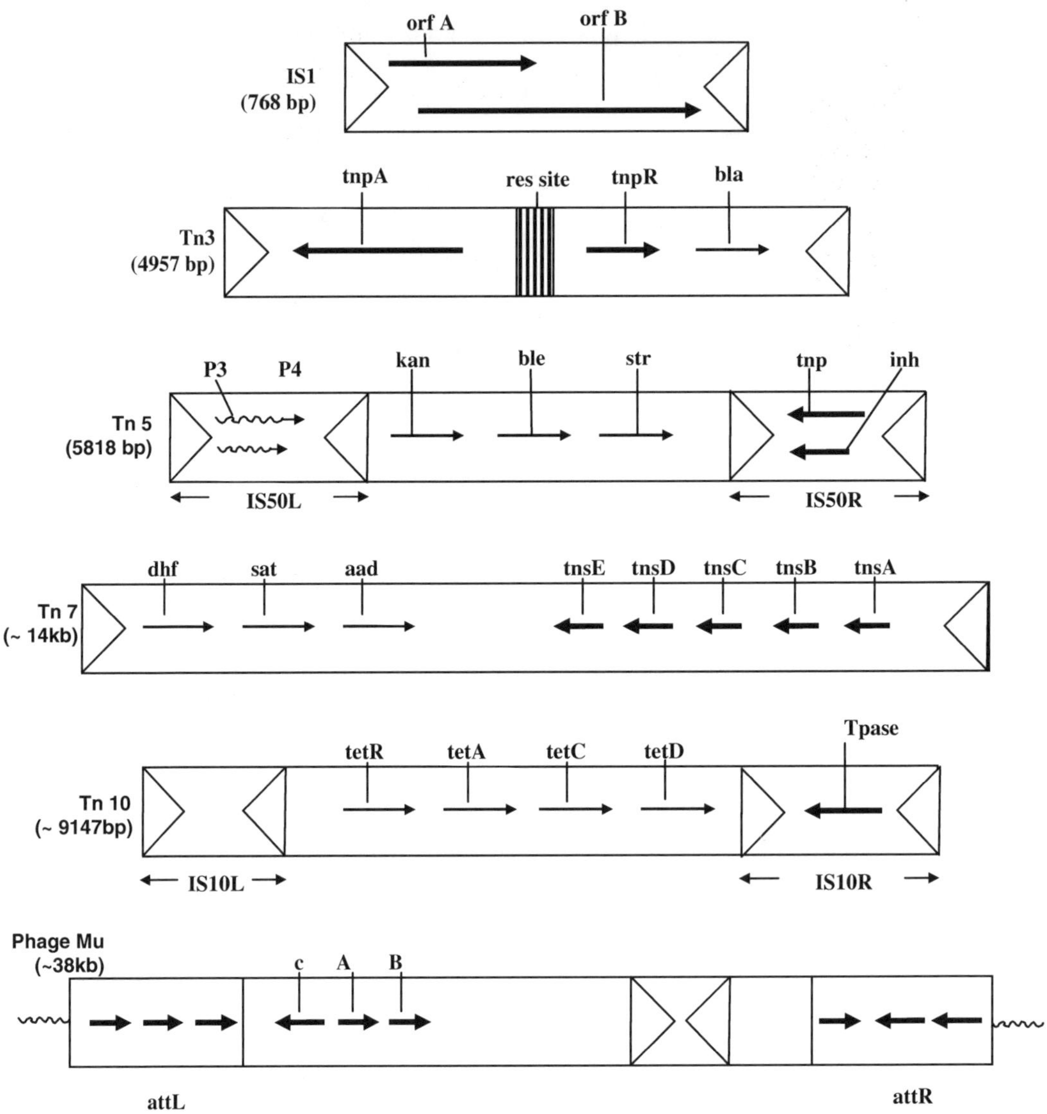

Fig. 7.2 Genetic organization of well-characterized TEs. *Thick arrows* represent functions associated with transposition (see text for details) and the thin arrows indicate other associated functions (Figure is not drawn to scale). *Triangles* denote repeat sequences. Composite transposons, Tn5 and Tn10, have been shown with their flanking insertion sequences, IS50, and IS10, respectively

Thus, several other methods of classification were adapted. For example, some closely related elements, showing some differences at the nucleotide sequence level are designated as isoforms and assigned to a family. Another system provides some information on the source of the element by putting the initials of the bacterial strain. For example, ISRm1 denotes its presence in *Rhizobium meliloti*. It will be no exaggeration to state that other systems of classification are likely to emerge, as newer and newer elements are getting discovered and described. While many of these have been found to be occurring independently, some such as IS1, IS10, and IS50 also occur as flanking sequences of Tn, such as Tn9, Tn10, and Tn5, respectively. They all share a structural plan even when they differ in their size and sequence (see Fig. 7.2).

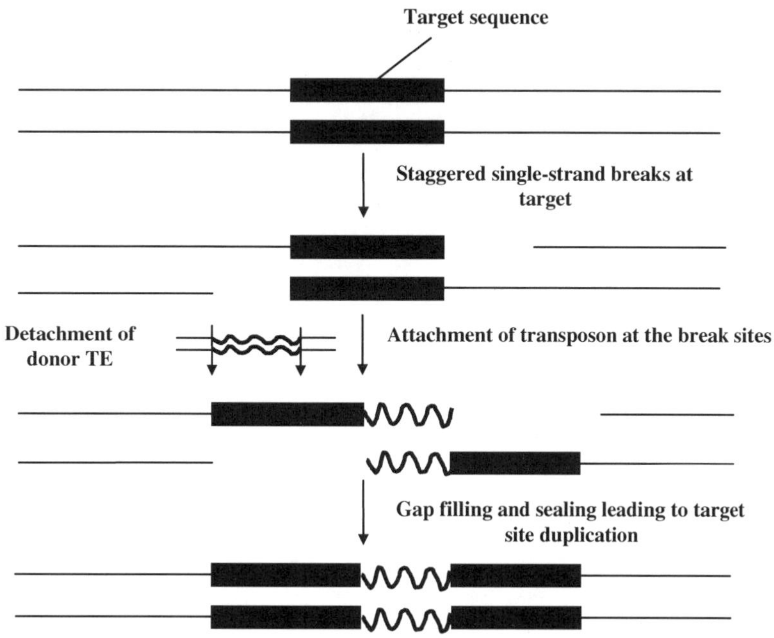

Fig. 7.3 Schematic diagram showing the generation of target site duplication (TSD). Transposon is released from a donor DNA molecule, and joins the target site cut in a staggered way. This is followed by gap filling. *Note* As the target site is fairly random, TSD sequences are different but they fall in distinct classes based on the size and is characteristic of a TE

This pertains to a sequence that codes for the enzyme, Tpase, and the terminal repeats of varying size. Some of the bigger members carry additional functions all geared toward the successful transposition.

IS were first identified by strong polar mutations in *gal* operon of *E. coli*. This was facilitated by the fact that most bacterial genes are organized into functional units called operon, which are regulated simultaneously by another set of genes. Moreover, many of these ISs carry strong termination signals, thereby disrupting not only the gene that they insert in but also other downstream genes of the operon.

7.3 Transposons

Besides IS elements, another group of TEs discovered in bacteria are called Tn. Unlike IS which do not code for any phenotypic marker, Tn may carry resistance to one or several antibiotics, metals, and toxins, in addition to those required for their mobility.

Their movement, therefore, can be traced easily (see Fig. 7.2). Because of this extra genetic information, Tns are much bigger in size ranging $\sim$5–10 kb. Several Tns have been identified till date and are labeled as Tn followed by a number, such as Tn1, Tn3, Tn5, Tn7, Tn9, Tn10, etc. These have been broadly classified into two categories: simple Tn (class II elements) and composite Tn (class I elements). (Table 7.2).

The basic difference between these two types is in the flanking inverted repeat sequences. While in the former (Tn3) they are short (TnA family), in the latter, such as Tn5, Tn9, Tn10, they are full-fledged IS elements. For example, in Tn5, it is IS50 L/R, in Tn9, IS1, and in Tn10 these IRs are IS10L/R. While in Tn5 and in Tn10, only one of the IS (IS50R and IS10R, respectively) is functional, IS1 flanking Tn9 and IS903 of Tn903 both ISs are functional. The flanking IS elements may be in either the same (Tn9) or more commonly, inverted orientation (Tn5, Tn10). In the former case, however, since IS element itself is flanked by short IRs, the composite element also ends in the short IRs. In cases where IS modules are closely related but not identical (Tn5, Tn10), the two ends are distinguished as L and R based on an arbitrary map of the transposon.

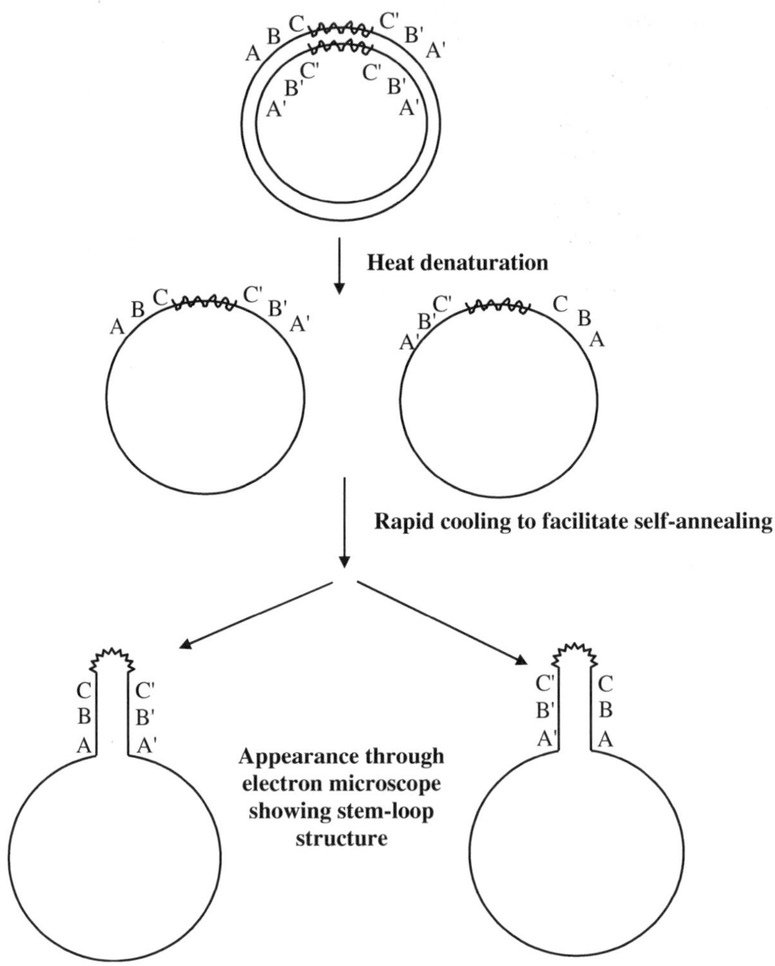

Fig. 7.4 Formation of a stem-loop structure by denaturation and intramolecular reannealing in a plasmid carrying a transposable element with inverted repeats. Both intact and nicked molecules are possible

Thus, if we take into consideration the organization of these Tn, they are no different from the IS elements. Like the latter, they are flanked by the inverted repeats of varying size, code for a Tpase and an in-built regulatory mechanism, and the rest of the length is devoted to antibiotic resistance or other properties (Table 7.2).

A composite transposon, therefore, may transpose as a unit or an active IS element at either end may also transpose independently. The guiding force behind the former event could be the strong selection advantage offered by Tn's marker genes, such as drug resistance. Composite Tn, thus appear to have evolved, when two copies of an IS element flanking a gene/(s) that provided a selective advantage got preferentially selected. The latter has the implication in providing the benefit to the host to survive and transmit these characters in a particular environmental niche.

7.4 Bacteriophage Mu

Another important class of bacterial TEs comprises temperate bacteriophages like Mu and D108. Phage Mu is a typical phage of ~38 kb that reproduces by transposition. As a phage, it carries all the genes for its growth and reproduction, but additionally also possesses genes *A* and *B* which code for a Tpase. The A protein is required for all transpositions and B is essential for replicative transposition events. Transposition also requires two ends of Mu, labeled *attL*

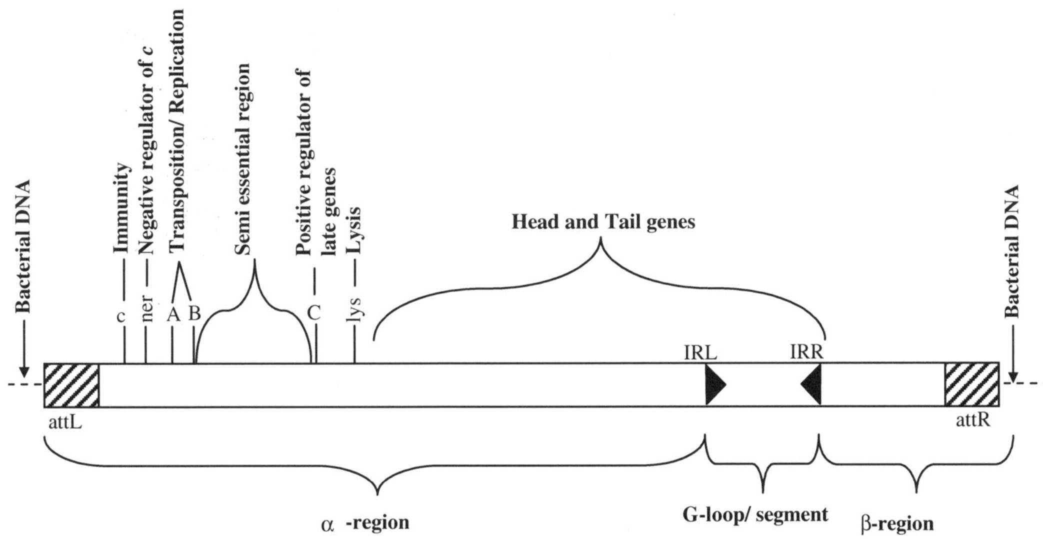

Fig. 7.5 Schematic representation of genome organization of Phage Mu (~38 kb) with only some genes shown here

Table 7.1 Properties of some common insertion elements (information from *E. coli*)

Element	Location and number of copies	Size (bp)
IS1	Chromosome 5 – 8	768
IS2	Chromosome 5, F plasmid-1	1327
IS3	Chromosome 5, F plasmid-2	1258
IS4	Chromosome, 1–2	1426
IS5	Not known	1195
Tn1000 ($\gamma\delta$)	Chromosome, 1 or more, F plasmid - 1	5980

Table 7.2 Properties of some common transposons

Element	Class (type)	Major resistance gene carried	Size (bp)	Terminal repeat size and their relative orientation
Tn3	II	Amp	~5000	38 bp IR
Tn5	I	Kan	5818	IS50, 1533 bp IR
Tn9	I	Cam	2638	IS1, 768 bp DR
Tn10	I	Tet	9300	IS10, 1239 bp IR
Tn1681	I	Ent	2088	IS1, 768 bp IR

IR inverted repeat, *DR* direct repeat, *Amp* ampicillin, *Kan* kanamycin, *Cam* chloramphenicol, *Tet* tetracycline, *Ent* enterotoxin

and *attR* (or MuL and MuR, respectively), which are not IR sequences (Fig.7.5). Phage Mu can insert highly randomly creating a 5 bp target site duplications (TSD). The phage is also highly efficient in introducing chromosomal aberrations.

While it would be virtually impossible to describe all the types of ISs and Tn, in this treatise we restrict ourselves to some of the common organizational and functional plans possessed by these elements. Readers interested in any specific element can easily get the required information from a large number of reviews and research papers available in the literature.

7.5 Organization

As described earlier, TEs share several features in their organization as depicted in Fig. 7.2. One special feature of the IS elements is that they are not only small but genetically compact. They generally code for no other function than those involved in their movement. This involves factors required in *cis*, that pertains to recombinationally active DNA sequences identifying the ends of the element, and also the enzyme, Tpase. Tpase is coded by one or sometimes two open reading frames (ORFs) that practically cover the entire length of the element. As described earlier, Tns also follow a very similar organizational plan. Besides coding for a Tpase, their structural complexity is devoted toward the additional genetic markers they carry and the genetic determinants for regulation of transposition pathway they sponsor.

7.5.1 Terminal Inverted Repeats

As described above, most TEs are flanked by repeat sequences. The ends of the majority of ISs are organized as short inverted-repeat sequences (IRs) ranging from 10 to 40 bp. The notable exceptions are some of the IS elements such as IS 91, IS 110, and IS 200/605 families which lack these IRs. In the well-characterized cases, the IRs can be divided into two functional domains. While domain II is involved in Tpase binding, domain I that actually includes the terminal 2 or 3 bp participates in the cleavage and strand transfer reactions during transposition. Many Tns are known to share similar organization but they may differ from IS in having multiple binding sites for Tpase. The IR sequence upstream of the Tpase gene, conventionally known as IRL, also carries the promoter for this gene. Such an arrangement may help in the autoregulation of Tpase synthesis, and also provide sites for binding of host-specific proteins. As described above, many Tns such as simple Tn are flanked by short IR sequences (e.g.,Tn3). In contrast, the IRs of composite Tn, you may recall, are in fact IS elements (e.g., Tn5, Tn9, and Tn10).

7.5.2 Transposase (Tpase or Tnp)

Tpase, as a multifunctional protein, is the most important component of a mobile element. It not only binds to the ends of an element but also catalyzes the movement of the same to another region of the genome or to another genome. A recent analysis has shown that genes encoding Tpase are the most prevalent genes in nature. This obviously reflects not only the most ubiquitous nature of the TEs, but also the very important role that this protein would have played in biological evolution.

From the well-characterized ISs, Tns, and phage Mu, a general pattern of functional organization of this protein can be revealed. In this organization, the sequence-specific DNA binding activities of the proteins are localized toward the N-terminal region and the catalytic domain is restricted toward the C-terminal end. This has been implicated in the interaction of the protein molecule with its target sequences that may regulate expression and activity. The C-terminal region of many Tpase appears to mask the DNA binding domain and serves an important purpose of regulating its activity. This also facilitates its *cis* activity, observed with many Tpase. Several other Tpase have the ability to generate multimeric forms essential for their activity. Based on the catalytic activity domain, Tpase as also the integrase (from retrotransposons) have been placed in the family of ribonuclease-like proteins, wherein the protein uses the same target site to catalyze both DNA cleavage as well as DNA strand transfer. These proteins, however, are active only when assembled in a synaptic complex called a transpososome. This complex provides a scaffold to support the transposition reactions and allows changes in its conformation so as to promote the different steps.

The distinct catalytic mechanism required for breakage and rejoining of DNA has been used to classify the Tpase into at least five distinct families. Basically, there are two pathways employed by these Tpases: while some cut/transfer/paste the original DNA (conservative pathway), others copy the DNA into a new target site (replicative pathway). The details of these pathways are described under mechanism of transposition.

7.5.2.1 DDE Transposases

These Tpases carry a conserved motif comprising a triad of amino acids: aspartate (D), aspartate (D), and glutamate (E) in its catalytic site. DDE motif not only functions in the co-ordination of a divalent metal ion binding but is also needed for the target capture. The members of this family employ cut-and-paste mechanism, and include phage Mu (first insertion), Tn5, Tn10, IS50, IS10, and eukaryotic Tn, such as maize Ac element, P element of *Drosophila*, and mariner.

7.5.2.2 Tyrosine (Y) Transposases

These also promote a cut-and-paste mechanism, wherein a site-specific tyrosine residue is involved in catalysis. The excised transposon forms a closed circle of DNA, which is then integrated into a new site and the original excision site is repaired. Examples are, Tn916 and DIRS1.

7.5.2.3 Serine (S) Transposases

They function similar to tyrosine Tpase except that a site-specific serine residue is involved in catalysis. Examples are, Tn5397 and IS607.

7.5.2.4 Rolling Circle (RC) or Y2 Transposases

These employ a copy-in mechanism by which a single strand is directly copied into the target site by DNA replication, employing host replication enzymes. Both old and new copies of Tn have one newly synthesized strand along with the older ones. Example: IS91 and others.

7.5.2.5 Reverse Transcriptase\Endonuclease (RT/En)

These are reported in retrotransposons, wherein an endonuclease is employed to nick the target site DNA, and the nick then serves as a primer for reverse transcription of an RNA copy by reverse transcriptase. Examples: LINE-1 and TP-retrotransposon.

It is clear that once more and more of these Tpases are characterized, we will learn more about their nature and function at molecular level.

7.6 Target Site Duplications (TSD)/Repeats

Another important feature of TEs is that upon insertion, most generate TSD or short direct repeat sequences (DRs) of the target DNA flanking the element. The length of the DR, but not the sequence, is fixed. The size ranges from 2 to 14 bp and is characteristic for a given element. Some ISs have been shown to generate DRs of atypical length at a low frequency and there may still be some that lack this property. The latter, however, could also be explained by homologous inter- or intramolecular recombination between two IS elements, each with a different DR or by the adjacent deletions arising from duplicative intra-molecular transpositions.

The mechanism of generation of DRs has been explained on the basis of attachment of each donor DNA strand of a cleaved transposon at the two ends of a target site cut in a staggered way. The subsequent gap filling, in fact, results in DRs (see Fig. 7.3).

7.7 Influence on Gene Expression

These elements were first identified by their inactivation (through insertion) of an active gene, suggesting thereby that they have the ability to influence the gene expression. One effect of this insertion could be on the transcription of gene/(s) located in the vicinity. Most often the transcriptional inhibition effects are due to the presence of termination signals or the polarity caused by the long untranslated stretches in mRNA of a TE, inactivating the gene/(s) located in the beginning of an operon. The degree of polarity is usually determined by the size of the transposable element, presence of internal transcription terminators, presence of promoter elements within the transposon or within the target operon, position of insertion relative to target gene or forming new promoter sequences at the insertional junction. Many of these latter events may even activate the expression of neighboring genes. Many IS elements such as IS1, IS2, IS5, and several others carry outwardly directed −35 promoter sequence in the terminal IRs. Thus, when the IS element gets inserted at a correct distance from resident −10 sequence, new promoters capable of driving the gene expression can be created. Similarly, an inwardly directed −10 sequence has also been detected in the IRL of several elements. If the two ends of such an element get juxtaposed by formation of head to tail dimers or due to the generation of circular copies of the IS, a combination of −10 with a resident −35 region in the neighboring right end can generate a relatively strong promoter. Many a time, this may lead to high expression of Tpase and a consequent increase in transposition activity (e.g., IS2, IS30, and IS901). With some elements, the expression of neighboring genes may also be effected by endogenous transcription escaping the IS and traversing the terminal IR (e.g., IS3, IS10, etc.). An interesting example has been reported from *E. coli,* where the otherwise cryptic, *bgl* operon is activated by insertion of IS1 or IS5. This has been explained by the change of the topology of the *bgl* promoter upon such an insertion.

We shall now discuss some of the better-known TEs to learn about their nature and functions, in detail.

7.8 IS1 Family

IS1, one of the smallest "autonomous element" was one of the first bacterial Insertion Sequences to be isolated and characterized. Originally isolated from F′ *lac-proB* plasmid and multiple drug resistance plasmid R100, it has now been reported from several other bacteria. Many variants of IS1, showing low or moderately high sequence divergence, have been placed in this family. IS1 is also a component of many compound Tn, such as Tn9 and Tn1681, where it is present in direct or inverted orientation flanking a gene for chloramphenicol acetyltransferase and heat-stable enterotoxin, respectively. Integration of IS1 is accompanied by a 9 bp target repeat though direct repeats of 8, 10, and 14 bp have also been reported. The element exhibits a strong preference for AT-rich target regions.

IS1 is 768 bp long that includes two ~ 23 bp imperfect inverted repeats, IRL and IRR, located at the two ends. In addition, two partly overlapping ORFs (*insA* and *insB′*) are also present in the 0 and −1 relative translational phases (Fig. 7.6). A transcript initiated from the promoter, P_{IRL}, partially located in IRL, is translated to give two products, InsA and InsAB′. While InsA encoded by *insA* frame is abundantly produced, InsAB′, the Tpase of this element arises from a programmed translational frame-shift between *insA* and *insB*. The site for this frame-shift is identified as A6C, though a modification of this motif (A7C or its replacement by GA2GA3C) leads to higher production of Tpase and consequently the increased transposition frequency. InsA binds specifically to two IRs, repressing transcription from *p*IRL, hence inhibiting transposition as well. Thus, the frequency of IS1 transposition is maintained at low levels.

7.9 Transposon Tn3

The classical example of simple Tn is TnA family (the best known members are Tn3 and gamma-delta element also called as Tn1000).

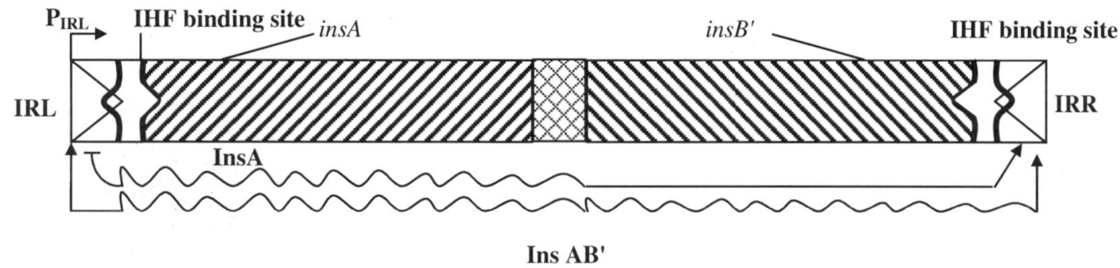

Fig. 7.6 Organization of IS1. It consists of two Inverted repeats (IRs), IRL (*left*) and IRR (*right*). IS1 promoter P_{IRL} is partially located in IRL, and two reading frames of InsA and InsB′ are shown with their overlapping region. Each terminal IR also carries an overlapping IHF site. InsA can bind to both the IRs and repress transcription from IRL. InsAB′ is the transposase (see also Fig. 7.2)

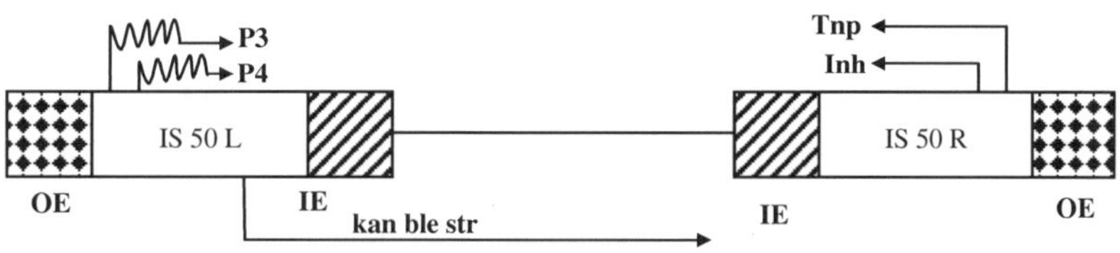

Fig. 7.7 Detailed structure of Tn5: two nearly identical IS50 elements flank a region coding for three antibiotic resistance genes. IS50R codes for both the transposase (Tnp) and an N-terminal truncated version of Tnp that acts as its inhibitor (Inh). IS50 L encodes c-terminal truncated inactive forms of both Tnp and Inh. Ends of both IS50 are identified by the 19 bp sequences, OE and IE, which are critical for Tnp binding (see also Fig. 7.2)

These are large (~5 kb) Tns flanked by short IRs, generally ~38 bp long, which are closely related. As they do not carry IS-type module, they are units comprising genes for transposition as well as for drug resistance (see Fig. 7.2). The transposition mechanism is replicative, and thus a co-integrate is formed as an intermediate. As shown in Fig. 7.2, of the two genes, *tnpA* codes for the Tpase and the product of *tnpR* functions as a transposition regulator. Tpase recognizes the ends of the element, as in IS, by binding to a sequence located in the 38 bp terminal IR and is further assisted by IHF such that the two bind co-operatively. The *tnpR* gene product has a dual function. It acts as a repressor and also provides the resolvase function. TnpR not only represses *tnpA* and thus regulates transposition, but also of its own. The *tnpA* and *tnpR* are divergently expressed from an inter-cistronic control region identified as *res*. Tn3 transposition takes place through a replicative pathway, wherein a co-integrate or a fused replicon between the donor and the recipient molecule is formed. TnpR, as a resolvase, is involved in recombination between the direct repeats of Tn3 so as to resolve the co-integrate into two individual replicons, each carrying a copy of the Tn (see Fig. 7.11).

7.10 Transposon Tn5

Tn5 is one of the most thoroughly studied and popularly used Tn. This composite transposon consists of a length of 5818 bp, wherein two nearly identical IS50 elements (1534 bp each and labeled as IS50R and IS50L) flank an intervening sequence coding for three antibiotic resistance for kanamyan, bleomycin, and streptomycin. IS50 elements placed in inverted orientation are defined by two 19 bp sequences referred to as outside end (OE) and inside end (IE), which differ at seven different positions but provide the critical Tpase binding sites. IS50R encodes the Tpase (Tnp) of 476 amino acid residues and N-terminal truncated version of Tnp inhibitor (Inh). Inh acts as *trans* dominant negative inhibitor of Tpase or Tnp. IS50L, on the other hand, encodes

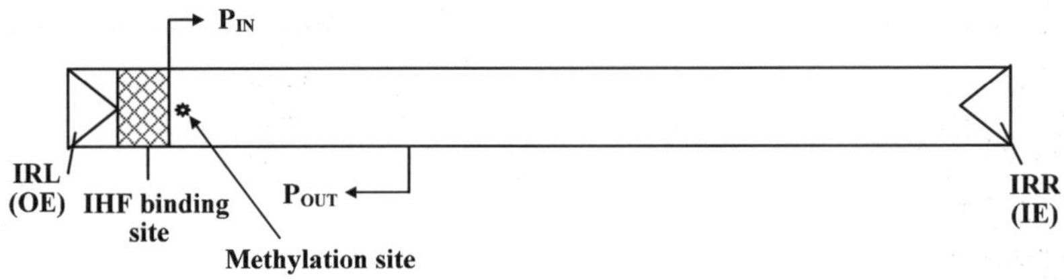

Fig. 7.8 Schematic representation of IS10R, one of the IRs flanking Tn10. The terminal IRL (OE) and IRR (IE) are shown as triangles. An IHF binding site lies next to IRL, followed by the promoter for TPase, P_{IN}. Dam methylation site (*) is also located there. The regulation of TPase expression is brought about by an anti-sense RNA transcribed from the promoter P_{OUT} (for other details see Fig 7.2)

C-terminal truncated, inactive version of both Tnp and Inh that makes the IS50L incapable of transposition (Fig. 7.7). The end sequences of IS50 are important for Tnp binding. The two ends come in close proximity as the Tnp bound to end sequences form a synaptic complex (see Fig. 7.11). In fact, for Tn5 transposition the end sequences are two inverted OEs, whereas for IS50 transposition they comprise inverted OE and IE sequences or two inverted IE sequences. Both IE and OE are suboptimal in their function and a hyperactive mosaic version (ME) has also been identified. The Tnp belongs to RNaseH superfamily of proteins, of which Mu Tpase (MuA), retroviral integrase, and RuvC are the other important members. The observation that Tn5 Tnp is essentially less active has gone in favor of both the host and the Tn. Hyperactive Tnp would lead to frequent transposition that may result in the genetic death of both the host and the Tn5. Transposition of Tn5 results in 9 bp target site duplication.

Since Tn5 is one of the most popularly used Tn, several variants have been generated for very specific uses in bacterial molecular genetics.

7.11 Transposon Tn10

Another well-characterized composite transposon is tetracycline resistance carrying Tn, Tn10. This 9147 bp Tn consists of a central region flanked by two non-identical ISs referred to as IS10L and IS10R (each 1329 bp long). Both IS10 and IS50 (as described above) belong to IS4 family of IS elements. Tetracycline resistance is mediated by four genes *terA, terC, terD,* and *terR* that carry out a drug/H^+ anti-porter function. IS10 carries 23 bp nearly perfect terminal IRs, called OE and IE, suggesting their relative position in Tn10 (Fig. 7.8). However, only OE has a binding site for the host protein, IHF. As with IS50, the IS10R represents the active module, transcribing Tpase from a long reading frame covering the whole length from a promoter referred to as pIN due to its orientation. A second promoter directed outwards, pOUT, is used for transcribing an anti-sense RNA which is complementary to region of Tpase mRNA carrying the ribosome binding site, thus helping in the regulation of Tpase gene expression. Insertion of Tn10/IS10 leads to 9 bp target duplication and requires a relatively well-defined target sequence with a symmetric 6 bp consensus: 5′ NGCTNAGCN-3′. Dam methylation sites exist in both IRL and IRR. Similarly, an IHF binding site that plays an important role in IS10 transposition has also been identified at the left end of IS10. Like Tn5, variants of Tn10 have been created for different applications.

Tn10 like Tn5 undergoes non-replicative, cut-and-paste transposition that is highly regulated. We shall discuss this aspect separately.

7.12 Transposon Tn7

Transposon Tn7 is a particularly sophisticated mobile element that has developed unique pathways to ensure its propagation. In one pathway, like many other TEs, Tn7 transposes to many different sites at low frequency showing

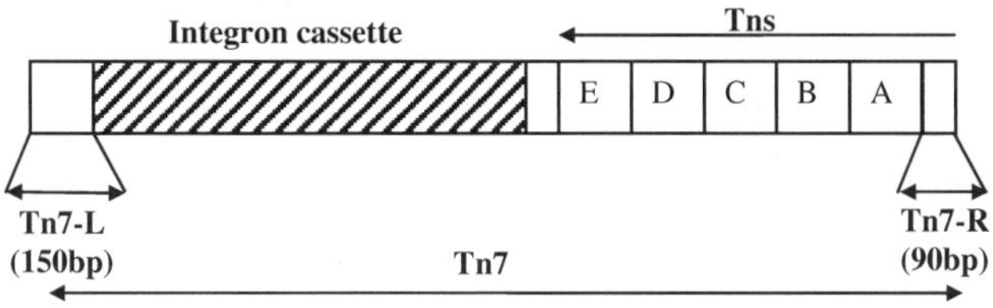

Fig. 7.9 Complete structure of Tn7. (for details refer text). This 14 kb transposon is flanked by small IRs of unequal sizes and carries five *tns* genes (A-E) required for transposition. Tn7 is associated with an integron that also confers the antibiotic resistance (see also Fig. 7.2)

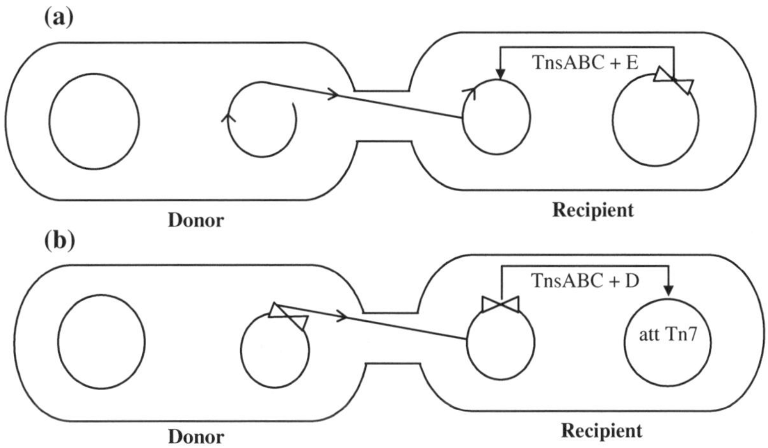

Fig. 7.10 Two ways of Tn7 transposition where TnsD and TnsE act as target site selectors. (**a**) Tns ABC+E pathway preferentially transposes a chromosomal copy onto a conjugal plasmid. (**b**) Tns ABC+D pathway promotes transposition into a single chromosomal site, *attTn7*

marked preference for broad host-range conjugal plasmids, whereas in the other it inserts at a high frequency into a single unique site in bacterial chromosome, referred to as *attTn7*. While the former pathway contributes to the dispersal of Tn7 among bacterial populations, the latter provides a safe haven as insertion into *attTn7* is not deleterious to the host.

Tn7 is a 14 kb long element carrying resistance for trimethoprim (*dhfr*), streptomycin (*sat*) and streptomycin/spectinomycin (*acdA*). These antibiotic resistance cassettes are part of an integron element (see later part of this chapter), located close to the left end of the Tn (Fig. 7.9). These cassettes are fixed in Tn7 owing to a mutation in the cognate recombinase but can undergo rearrangements in other hosts that express a related recombinase, leading to alternative combinations of antibiotic resistance or even other genes.

The ends of Tn7, at which transposition machinery acts, consists of $\sim$150 bp Tn7L and $\sim$90 bp Tn7R lying at the conventional left and right ends, respectively. Although the end sequences are related they are not identical facilitating a preferential orientation at particular targets. Each end of Tn7 contains a series of 22 bp sites, three non-overlapping sites in Tn7L and four overlapping sites in Tn7R, as Tpase binding sites. The Tn7 transposition machinery consists of five proteins TnsA, TnsB, TnsC, TnsD, and TnsE. The genes for these are located at the right end with their 5′ termini closest to the Tn7R (see Fig. 7.2). Of these, TnsA and TnsB together form the Tpase that specifically recognizes the ends of the transposon. TnsAB excises Tn7 from the donor site by introducing double-stranded breaks at each end, and then joins the exposed ends to the target DNA resulting in 5 bp target site duplication.

The proteins that make up the Tpase are quite different. TnsB, which carries out the chemical step at the 3′ end of Tn7 is a member of the retroviral integrase superfamily. TnsB also has the ability to recognise the cis-acting ends of Tn7 and has binding sites, as mentioned above, for this purpose. TnsA, on the other hand, which executes the breaks at the 5′ end of Tn7, resembles a type II restriction endonuclease. TnsA does not have a detectable DNA binding activity and thus relies on TnsB to recognize and cleave at the ends of Tn7. TnsC is an ATP-hydrolyzing protein which promotes excision and insertion of Tn7, interacting both with target DNA and TnsAB. TnsD and TnsE are alternative target selectors while also acting as activators of core TnsABC machinery (Fig. 7.10). Tns ABC+E transposition shows an interesting preference for conjugal plasmid relying on specialized DNA replication occurring during conjugal plasmid transfer. Low frequency transposition through this machinery could also take place in the bacterial chromosome, but such insertions show a preference for the region where bi-directional DNA replication terminates.

Tns ABC+D, on the other hand promotes Tn7 insertion into a specific DNA sequence called *attTn7*, which is found in many bacteria. TnsD acts as a sequence-specific DNA binding protein. The sequence recognized by TnsD lies at the 3′ end of the bacterial glutamine synthetase gene, *glmS*. The *glmS* gene is highly conserved and plays a role in the synthesis of N-acetyl glucosamine. The actual point of Tn7 insertion lies downstream of *glmS* gene, thus not affecting the gene and hence, the bacterial host as such. This appears to be a smart strategy such that Tn7 finds a safe insertion site in diverse hosts. Thus, while Tns ABC+E promote horizontal transfer of Tn7 between diverse bacteria, TnsABC+D pathway would facilitate vertical transmission to daughter cells in a host. The sequence-specific binding of TnsD, in fact, facilitates the recruitment of TnsC, the Tpase TnsAB, and the ends of the transposon for target site recombination. Tns ABC+D - mediated transposition is a complex phenomenon not only requiring multiple proteins but also host factors. Many Tn7-like elements have also been discovered with a similar insertion site that should provide additional information about the spread of Tn7 and its associated mechanism.

7.13 Bacteriophage Mu

One of the interesting class of TEs in bacteria is a temperate phage, Mu. In fact, it was described as a mutator phage, hence called Mu by Taylor in 1963. This nomenclature was based on high frequency of mutations generated by the lysogenization of the phage. While Mu is one of the best-known TEs in Gram-negative bacteria, several Mu-like bacteriophages are now known. Phage Mu is a giant transposon, ~38 kb in length that integrates into host *E. coli* genome, both during lytic and lysogenic cycles.

Being a bacteriophage, its organization is very different as it must carry the genetic components for its own multiplication. However, it also codes for a Tpase which acts at the two ends of the genome conventionally referred to as *attL* (MuL) and *attR* (MuR) (see Figs. 7.2 and 7.5), but unlike other TEs, these ends are not inverted repeats. Each end of Mu has three Tpase (MuA) recognition sites, though not all are essential for transposition. Two IRs (IRL and IRR) are found located within the genome flanking the G-segment carrying the genes coding for tail fiber proteins and help in controlling the host range of the phage. Two gene products required for the transposition are the MuA or the Tpase, and MuB, the accessory protein required for replication and full transpositional activity. Upon infection of a host cell, Mu genome integrates randomly into host genome by a conservative, cut-and-paste mechanism. However, when Mu replicates, it undergoes repeated cycles of transposition that follows a replicative pathway. Mu transposition results in 5 bp target sequence duplication and is often accompanied by large-scale genomic rearrangements.

7.14 Delivery of Transposable Elements

Looking at the plethora of applications, TEs have served as valuable tool in molecular genetic analysis of bacteria. To fulfill such a requirement, both in vivo and in vitro transposition systems have been developed. This requires a careful consideration of the strategy used for the delivery of TEs. Generally speaking, such a strategy depends upon the target strain and whether the target is the bacterial chromosome, a phage, or a plasmid. For chromosomal targets, the pathways utilized are those mediating intercellular gene transfer as shown in Fig. 7.1. In other words, the Tn is usually transferred via a phage, a suicide vector, or an F-plasmid. Of these, the suicide vectors that are unable to replicate in a new host and a recipient that cannot grow under the special conditions (presence of an antibiotic to which the Tn is resistant) are chosen. Bacteriophages are the most convenient type of the delivery vehicle when insertion in the bacterial chromosome is required. The conditions used are the ones that disable the phage genome replication and the cell lysis, thus favoring stable insertions. Phage lambda vehicles are often used to isolate Tn10, Tn5, and Mu insertions. The latter, of course, can also be used directly by phage infection pathway. Similarly, conjugative multicopy plasmids and bacteriophage-based vectors have been constructed for specific use. When the target molecule is a phage or a conjugative plasmid, any type of the replicon other than the target itself can be utilized.

To meet the specific applications and achieve stable insertions, mini-transposon constructs are very useful. These mini-transposons are curtailed versions of their wild-type counterparts and do not contain a Tpase gene within their boundaries. While the latter make them specifically controllable, these mini variants can be tagged with genes having a selectable phenotype (antibiotic resistance) or with a reporter gene (*lacZ, phoA,* etc.). Such elements have found wide applicability, which besides insertional mutagenesis may consist of gene fusion (both transcriptional and translational), promoter fusion, in vivo cloning, functional analysis, and many others.

7.15 Mechanisms of Transposition

In spite of the large variety of TEs, another common feature is that their transposition mechanism follows one or the other pathway as shown in Fig. 7.11. One is called the conservative or cut-and-paste mechanism while the other is typically non-conservative or replicative pathway (Fig. 7.11). Conventionally, the conservative pathway comprises the excision of the element from a donor site/molecule and its integration into a fresh site. The replicative pathway involves the intervening step of replication that generates a co-integrate (fused replicon) from where the two copies are resolved and distributed to both the donor and the recipient as depicted in (Fig. 7.11). Another common feature of the transposition reaction is that it requires the formation of a high order protein-DNA complex called transpososome. We shall discuss these pathways taking the well-known examples from both these categories.

7.15.1 Conservative Transposition

The two classical examples of this category are Tn5 (IS50) and Tn10 (IS10).

Tn5 is an excellent model system for understanding the molecular basis of DNA-mediated transposition. Although DNA transposition is a complex phenomenon it is surprising that it requires only three macromolecules. Two of these comprise the transposon DNA (especially the ends) and the 476 amino acid long Tpase (Tnp) that catalyzes the transposition. The third constituent is the target DNA sequence that may be located on the same or on a different DNA molecule, with little or no specificity in the target sequence.

As shown in Fig. 7.11, the first step begins at the donor site, where two molecules of Tnp are bound in a *cis/trans* relationship with two 19 bp end sequences, forming a synaptic complex. It is also known that the DNA is distorted at the transposon-donor DNA junction that facilitates the subsequent catalytic steps. The latter results in DNA cleavage through 3 steps: 3′ strand nicking, hairpin formation, and hairpin cleavage. The activated oxygen serves as nucleophile to nick the 3′ strand at both the ends, in a reaction in which Mg^{2+} ions play an important role. This first reaction is shared by all DNA Tpases,

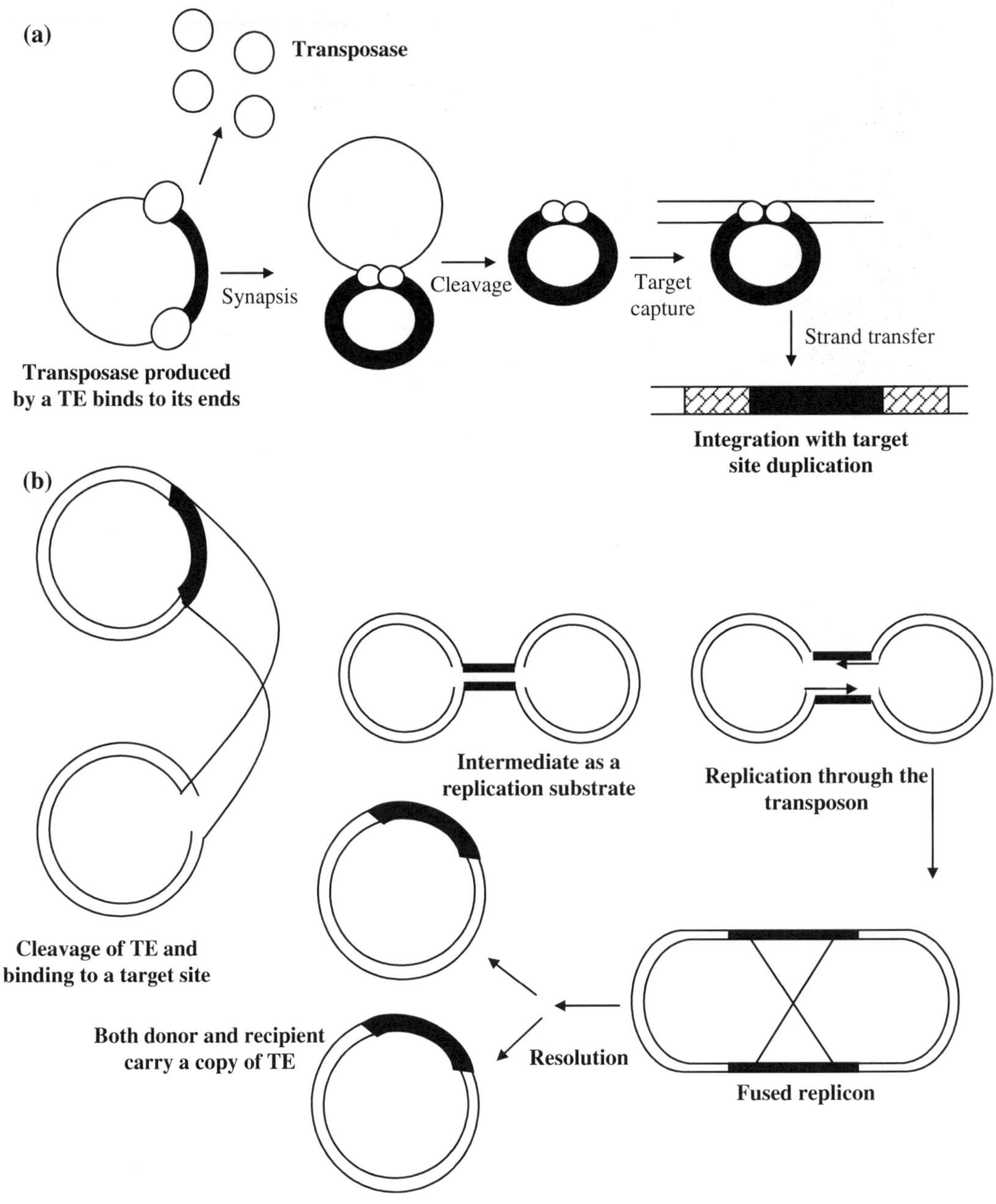

Fig. 7.11 Two pathways of transposition: (**a**) Conservative or cut-and-paste mechanism, wherein the transposase (OO) binds to the ends of the transposon (thick dark region) and brings these ends together in a synaptic complex. Transposon is released by strand cleavage and captures a new target site. (**b**) Replicative transposition, where the transposon released by strand cleavage gets attached to a new site or target and generates a replication intermediate. This leads to the formation of a fused replicon wherein the two copies of the transposon are resolved and get distributed to the donor as well as to the recipient molecule

retroviral integrases, and the long LTR retrotransposases. Retroviral integrases also share structural features, especially the active site with Tn5 Tnp. In the second step, the nucleophilic attack of the 3′ OH group of the transferred strand on to 5′ (non-transferred)-strand results in a DNA hairpin. Hairpin generation thus detaches the transposon from the donor backbone, a feature shared with the Tn10 system. The hairpin structure is finally nicked suggesting that some movement of the non-transferred (5′) strand must take place to facilitate target DNA docking. Hairpin resolution is important to re-expose a 3′ OH at each of the transposon terminus. The next step involves target DNA capture, where some sequence bias may exist. Subsequent to this, the two 3′ OH ends attack the target DNA phosphates 9 bp apart on the opposite strand (strand transfer) resulting in the integration of the transposon. The strand transfer complex is finally disengaged and the host fills the two 9 bp gaps at either end of the integrated transposon to generate the DRs.

IS10 (Tn10) transposition also involves double-strand cleavage at the ends of the IS by Tpase binding, separating it from the donor backbone. The outside end of IS10 carries the IHF binding site and a synaptic complex is formed between the two ends of the transposition substrate, Tpase and IHF, forming a transpososome. Subsequent steps are very similar to IS50 (Tn5). One important difference, however, is that the target site selection by Tn10 is highly sequence-specific. The fully excised transpososome initially binds weakly to a target DNA and only upon encountering a preferred target sequence does it form a more stable target capture complex. Another host protein, H-NS acts as a positive regulator of Tn10 transposition by inhibiting intramolecular transposition.

7.15.2 Replicative Transposition

This is another mode of mobility exemplified by the members of TnA family and the phage Mu. The former, you may recall, are Tns (~5 kb) that do not carry IS-type flanking transposition modules. They consist of genes essential for transposition as well as for drug resistance (ampicillin) bracketed by closely related inverted repeats (TIR), generally ~38 bp in length. TnA family includes several related Tn, of which Tn3 and Tn1000 (gamma-delta) are the best-known members. The two genes required for replicative transposition are *tnpA* coding for the Tpase and *tnpR*, the product of the latter acts as a repressor as well as a resolvase, as described earlier. Also, the transposition requires the two ends, like other TEs.

Tpase, TnpA binds to a sequence of ~25 bp located within the 38 bp TIR. Close to the TnpA binding site is a site for binding of the host protein, IHF. The two bind co-operatively to these sites. TnpA not only recognizes the ends of TE but also makes a 5 bp staggered cut in the target DNA. The cleavage of the transposon as well as the target DNA takes place at 3′ end. This is followed by the strand transfer, wherein the 3′ OH end of the cleaved transposon is joined asymmetrically to cleaved target DNA. This generates a replicational substrate, such that host DNA polymerase replicates across this structure to produce what is known as co-integrate or replicon fusion. This structure contains two copies of the Tn at the junction points, each containing one original DNA strand and one newly synthesized strand. This co-integrate is finally resolved into two molecules by a resolvase as described below, generating both donor and recipient DNA molecules carrying a copy each of the Tn (Fig. 7.11). Mutations in *tnpR* increase the transposition frequency confirming the role of repressor coded by this gene. The *tnpA* and *tnpR* are expressed divergently from an A-T rich intercistronic control region. TnpR or resolvase is a site-specific recombinase that resolves the direct repeats of Tn3 in a co-integrate. The site of resolution or *res* is made of three sites each about 30–40 bp long. They share sequence homology in terms of consensus sequence with dyad symmetry and TnpR binding that occurs independently at these sites. Site I includes *res* and overlaps with the start point of *tnpA* transcription. Site II overlaps with the start point of *tnpR* transcription. Binding at all the sites may be essential for holding the DNA in an appropriate topology.

During resolution, six resolvase dimers bind to two *res* sites forming a structure called synaptosome. Moreover, DNA is bent at the *res* site by binding of TnpA. Resolution does not involve any input of energy and relies on breaking and rejoining of bonds. Cleavage is accomplished by trans-esterification reaction with a resolvase subunit linked to 5′ P of the target bond at site I. In other words, resolvase is covalently attached to 5′ ends of the two double-stranded cuts made at the *res* site. Cleavage occurs symmetrically at a short palindromic region to generate two base extensions. The reaction resembles the action of λ Int at *att* site and is further seen by the identical base sequence of 15–20 bp of *res* with that at *att*.

IS1 could generate both simple insertions and replicon fusions (co-integrates). The latter like other replicative Tn consists of two copies of IS1 in direct repeat orientation. During transposition of IS1-derived Tn, excised circular copies have also been detected as in the case of IS911 (of IS3 family). Such forms may integrate into the target, giving rise to simple insertions. The replicative transposition, therefore, could belong to an intramolecular event. High levels of Tpase (InsAB′) in the presence of suitable IS1 ends induce the host SOS response, suggesting that IS1 Tpase possesses endonucleolytic activity. InsAB′ has an amino acid triad characteristic of the phage λ family of integrases.

One of the classical examples of replicative transposition, and the one that has been most thoroughly studied, is represented by phage Mu. Transposition of bacteriophage Mu uses two cleavage sites and six Tpase recognition sites at each end of the Mu genome. Named L1, L2, L3, at the left end, and R1, R2, and R3 at the right end, these sites share a 22 nucleotide consensus sequence with no apparent internal symmetry. All the binding sites are not equally important for transposition. Like many other Tns, transposition of Mu occurs within complexes called transpososomes. Transpososomes contain multiple subunits of a Tpase protein bound to DNA sequence at both the ends of Mu. These protein-DNA complexes are also called "synaptic complexes" because they bring the transposon ends together. In the transpososome, at least two sequential chemical reactions take place:

(i) DNA cleavage, introduced precisely at the end of the transposon, and
(ii) DNA strand transfer, wherein the nicked stand is joined to a target DNA.

While the reaction sites on the transposon or "donor" DNA are sequence-specific, the target sites of Mu are not, though the DNA regions close to the Mu end sequences are poor targets for transposition. Phage Mu Tpase, MuA is monomeric but forms a tetramer when it binds to the recognition sites. MuA specifically binds by its N-terminal domain bringing two MuA subunits close together, allosterically activating the Tpase. The recognition sites also help define the cleavage site that is contacted by the catalytic domain of MuA on the opposite strand. Thus, MuA subunits bound to the R1 site engages the left end cleavage sites and the subunit bound at L1 engages the right end cleavage site. The cleaved phosphodiester bond at the end of the Tn is separated from the outermost recognition sites (R1 or L1) by five base pairs. The last two base pairs have been identified to be 5′ CA (or 5′ TG complement). This particular sequence, TG/CA marks the cleavage sites of many distantly related TEs, including retroviruses and bacterial ISs. Besides the terminal Tpase binding sites an internal activator sequence (IAS or transpositional enhancer) is required for efficient transposition. Lying close to the left end, IAS contains an IHF binding site and overlaps with two transcriptional operators, which are binding sites for Mu repressor.

MuB, an accessory protein in transposition, is an ATP-dependent DNA binding protein that in collaboration with MuA promotes efficient replicative transposition to new target sites. MuB binds target DNA, delivers this target segment to Tpase, and activates Tpase's catalytic functions. MuB is known to activate the tranpososome for DNA strand transfer, a prerequisite for its integration. Thus, MuB-bound DNA acts as an efficient target for Mu Tn. In the presence of ATP, polymeric MuB clusters are formed at A/T-rich sequences in the DNA coinciding with the transposition hotspots. When MuA binds to the end sequences, it stimulates the MuB ATPase triggering MuB dissociation from neighboring DNA. Other components of successful transposition are host-encoded DNA bending proteins, HU and IHF, and highly supercoiled donor DNA. All steps in Mu transposition pathway have been worked out. The product is a branched molecule (the Shapiro intermediate) in which the donor DNA is covalently linked to the target via 5-nt single-stranded gap. This strand transfer (ST) product then serves as a template for Mu DNA replication, which leads to the formation of co-integrates. Simple insertions, on the other hand, are formed by repair of the ST product. The host proteins repair fill the gap generating a 5 bp TSD flanking the inserted Mu. The co-integrate is further resolved by homologous recombination between the two copies of Mu.

7.16 Regulation of Transposition

Although the TEs are best characterized by their mobility, their movement does not go unchecked. As the transposition could lead to the loss of a gene function with far-reaching consequences, each transposable element in conjunction with their host has evolved a mechanism to regulate the level of transposition.

IS1, codes for two products InsA and InsAB′ of which the latter is the Tpase, the transcription is initiated from *p*IRL located partially in IRL. The small, basic but abundantly produced InsA acts as a transcriptional repressor thus limiting the amount of Tpase. It could also bind to the ends of IS1 and thus compete with Tpase for transposition. The production of InsAB′ takes place through a translation frame-shift in the region A6C. A modification of this region to A7C on GA2GA3C leads to constitutive production. This suggests that there is an inherent mechanism to regulate this frame-shift resulting into low frequency of natural transposition ($\sim 10^{-7}$). There are also suggestions that a translational restart may have to take place within InsA frame to give rise to Tpase, which may not be very efficient. Transposition activity depends upon InsAB′/InsA ratio, ensuring that IS1 is not activated to high levels. A Rho-dependent transcription terminator-like region identified at the end of *insA* may also limit the production of InsAB′.

The Tpase of ISI0 (= Tn10) is expressed from a single long reading frame from the promoter P_{IN}. The expression is regulated by external transcription by an IR sequence located close to the left end by a process called sequestration of translation initiation signals. In the P_{IN} promoter, there is a site for Dam methylase in the −10 region and methylation is known to reduce its activity. A second control is exercised by a small anti-sense RNA transcribed from outwardly directed promoter P_{OUT} located proximal to IRL. This RNA is complementary to the region of TPase mRNA carrying the ribosome binding sites. This can interact with mRNA in *trans* and inhibit its translation (see Fig.7.8). IRR contains other Dam methylation sites as well. When IRL and IRR sites are methylated the activity at the ends of the Tn and consequently the transposition is reduced.

IS50 (= Tn5) has evolved elaborate series of mechanisms for regulating its movement and shares many of these features with IS10. More specifically, it codes for two proteins, its Tpase called Tnp and an inhibitor, Inh. Both are expressed from a single long reading frame using two promoters P1 and P2, respectively. Inh is translated using an alternative initiation codon and lacks the N-terminal 55 amino acids. Like IS10, Tnp expression is protected by sequestration of translation initiation signals as well as Dam methylation, the sites for which are found in P1, the endogenous Tnp promoter.

The ends of IS50, OE, and IE carry a series of sites modulating Tnp expression. Besides the binding sites for Tnp (19 bp end sequences), OE carries binding sites for host DnaA protein, and transposition activity is reduced in *dnaA* host. IE, on the other hand, carries binding sites for the host protein Fis and three consecutive Dam methylation sites. Both can bring down the transposition activity. Inhibitory action of Inh involves the formation of hetero-multimers between Inh and Tnp that are inactive. Interestingly, while Inh can function in *trans*, Tnp like many Tpases preferentially acts in *cis*. Both OE and IE function at suboptimal level and thus Tnp is essentially less effective.

Mu Tn is also subjected to several types of regulation. Mu transpososome engages multiple DNA sites that consist of recognition sites and cleavage sites. In addition, DNA topology, the requirement of divalent cations, and various structural transitions, all contribute to the regulation of transposition. MuA favors certain consensus sequences as this contributes towards efficient transposition. The multiple and non-symmetrical Tpase binding sites can provide a functional distinction between the ends either in the assembly or activity of the synaptic complex. Impinging transcription that is the simple transcription across the end of the element, more so the left end reduces the transposition activity of phage Mu.

Transposition activity of several Tn including phage Mu can be modulated by several host factors. They include DNA architectural proteins IHF, HU, Fis, and H-NS. Phage Mu has a binding site for IHF approximately 1 kb from the left end. The severe DNA bending introduced by this binding regulates a divergent promoter that modulates the repressor and Tpase expression. Synergistically acting dimeric protein HU presumably facilitates transpososome assembly during Mu transposition.

The level of DNA supercoiling clearly affects Mu transposition. Phage Mu also carries a strong gyrase site located approximately in the middle of its genome. It is believed that gyrase-driven supercoiling promotes end pairing and assembly of the transpososome.

Another strategy that assures that the transposition occurs under appropriate circumstances is based on the requirement for specific target sequences, best exemplified by Tn7. Tn7 does not shed its flanking donor DNA unless a nucleoprotein complex, containing the specific target site, *att Tn7*, is assembled.

While we have discussed some very specific examples, there can be several other ways by which Tpase/transposition can be regulated. In the succeeding paragraphs we shall get an overview of these mechanisms.

Many a time, the endogenous promoters driving Tpase expression are not very efficient and limit the Tpase levels. Very often such promoters are partially located in the terminal IR that allows regulation by binding of Tpase or its truncated derivatives.

Many IS elements may transpose creating a right disposition of IRL −10 hexamer, which with a −35 hexamer resident in the neighboring IRR may generate a relatively strong promoter *Pjunc*. Although this can lead to high expression of Tpase, the latter's binding to this junction not only represses further expression but also results in cleavage at both ends that leads to disassembly of the *Pjunc*. Regulation may also occur by Dam methylation of sites close to or overlapping the endogenous promoters. As discussed earlier, this has been found to be the case in IS10, IS50, and IS903; when fully methylated the promoter is inefficient.

The Tpase ORF of several IS elements (IS1, IS2, IS3, IS5, IS30) carry sequences resembling transcription termination signals. For IS1 this is partially Rho-dependent.

As discussed earlier, transcription across the end of the element (impinging transcription) has been reported to reduce transposition activity of both phage Mu and IS1. This could be due to the disruption of the complexes between Tpase and the cognate ends. Stability of Tpase mRNA could also influence both the level and type of the translational product. Many ISs such as IS10 and IS50 control translation by sequestering translation initiation signals in an RNA secondary structure. Another way of translational control is exerted by anti-sense RNA, as in case of IS10 (Tn10), an anti-transcript from the promoter P_{OUT}, located within the Tpase ORF and transcribed in the opposite orientation, is produced. This locks out the translation initiation signal at the 5′end of Tpase messenger, effectively downregulating the same and consequently the transposition (see Fig. 7.8). Some IS elements carry a strong promoter (IS186) but the transcripts do not have an obvious ribosome-binding site (RBS). The translational rate of these leaderless RNA is obviously very low.

We have earlier discussed the translational frame-shifting in case of IS1 which is responsible for the production of a functional Tpase. The frame-shifting frequency is a critical determinant of transposition activity. The members of IS21 family have two consecutive reading frames that may share overlapping or have closely placed termination and initiation codons. IS21, for example, contains two translation initiation codons lying eight codons apart. The use of these alternate codons is under physiological regulation.

The translation termination codon of many Tpase genes resides in their IR sequences and couples the termination with Tpase binding and transposition activity. There are other IS elements that do not possess a termination codon (IS240C, IS427, etc.) and termination would presumably occur outside the element. Yet another group (IS870, ISRf1) generates a termination codon when they insert in a specific target sequence. Another level of regulation can be due to post-translational modifications such as proteolysis. As discussed before, the N-terminal region of Tpase may regulate the activity of the full length protein either by competing with Tpase for binding (IS1) or by forming hetero-multimeric complexes (IS911). Transposition of many such as IS1, IS30, IS911, and Tn3, shows strong temperature sensitivity and thus an optimal temperature is a prime requirement.

In addition, regulation may also operate at the level of correct assembly of the ends and the Tpase in a transpososome. Thus, any non-productive cleavage or strand transfer will be limited. While transcriptional control can be an important aspect the elements with multiple and non-symmetrical Tpase binding sites at their ends (Mu, Tn7, and IS21) are examples of such a mechanism. A similar effect can also be produced by the architecture of the transpososome (IS50 and Mu).

Requirement for a specific target site selection or preferred target choice may also translate into a regulated transposition. Similarly, many Tpases show preferential *cis*-activity (IS1, IS911, and IS50) that is, Tpase acts more efficiently on the element from which it is produced. This may prevent activation by one element of the others in the same cell that otherwise will prove highly detrimental to the cell.

The host cell or the factors play an important role in regulating the transposition. The histone-like or DNA chaperones IHF, HU, Fis, and H-NS have all been implicated. These can lead to several changes such as DNA bending, stimulating the repressor binding, or they may facilitate transpososome assembly. Both Fis and DnaA, and H-NS influence IS50 and IS1 transposition, respectively, by regulating transpososome assembly. Similar functions have been assigned to acyl carrier protein (ACP) in case of Tn3, whereas in Tn7, ACP along with ribosomal protein L29 is implicated in increasing the binding of TnsD protein.

Another important host factor is DNA supercoiling (as in Mu) and thus the influence of gyrase (Mu) and topoisomerase I (Tn5) can be observed. Dam methylation has been implicated in the transposition of several elements as discussed earlier. Some protease/chaperone systems have also been identified. ClpX, a molecular chaperone with ATPase activity and often associated with ClpP protease, plays an important role in Mu transposition .

7.17 Transposition Immunity

Another important property of some TEs is the transposition immunity wherein a target molecule carrying an element prevents the transposition of another copy within or to another site close to the previous one. Transposon Tn10 is virtually an exception and does not exhibit transposition immunity. Although characterized only for example, for TEs, such as Mu, Tn7, and Tn3, some vestiges have also been detected in IS21. Generally speaking, IS elements have not been observed to adopt this strategy, but there may be another pathway through which the transposition will be regulated. Moreover, such an immunity may have high relevance, looking at the detrimental effects of transposition. In bacteriophage Mu, transposition immunity is displayed by target DNA carrying Mu ends and is operationalized by MuB protein. It may be recalled that MuB plays a key role in target capture and strand transfer by binding to the DNA ends that provides a preferential target for the MuA, Tpase complexed with Mu ends, as also stimulating the Tpase activity. Although MuB displays an ATPase activity that is facilitated by MuA and the target DNA, it is ATP and not ATP hydrolysis that is required by MuB to act. MuA bound to an immune target (TE ends) interacts with MuB and reduces the utility of the same target site for transposition. In other words, MuB gets redistributed to DNA sites that do not contain MuA binding sites (or transposon). Mu transposition immunity also senses the domain structure of bacterial chromosome or its own supercoiling.

In Tn7, TnsB which binds specifically to the ends of the Tn and TnsC, the ATP-dependent DNA binding protein acts as molecular switch to impose immunity on target DNA containing Tn7. The target DNA containing Tn7 ends is also bound by TnsC and is subsequently inactivated by TnsB, for which protein–protein interaction is responsible. The target DNA promotes these interactions by tethering TnsB and TnsC in high local concentrations. TnsB appears to trigger the dissociation of TnsC from the Tn7 end-containing target DNA. These interactions (TnsB and TnsC) not only prevent Tn7 from inserting into itself but may also facilitate, if you recall, the preferred site selection that is the hallmark of Tn7 transposition.

Transposition immunity has also been observed for Tn3 and the related Tn1000 ($\gamma\delta$). In Tn3, only the 38 bp terminal IR is required along with its Tpase (TnpA). As in the other systems, immunity is mediated by Tpase binding to its ends, and thus functions in *cis*. IHF which stimulates Tpase binding to the IR of Tn1000 also influences immunity.

7.18 Target Site Selection

The TEs were first identified by their mutational effect which appeared to be highly random. This led to the proposition that the elements insert without any target site selection specificity, and thus can be exploited for the generation of a whole variety of mutations. Upon subsequent characterization, however, it became clear that many of them may have a localized preference for a target to some which may exhibit a strong target site choice. Once again, we shall discuss this aspect taking some of the well-characterized examples. On one side of the spectrum is phage Mu, whose target sites are not very sequence-specific, strengthening the fact that Mu can produce a whole variety of insertional mutations. Also classified in this category is transposon Tn5 (or IS50) which inserts without notable target sequence specificity. In contrast, Tn10 (or IS10) insertion occurs in a relatively well-defined target sequence carrying a symmetric 6 bp consensus sequence: 5′NGCTNAGCN3′. IS1 (and, therefore also theTn9) strongly prefers AT-rich regions. The most notable exception as far as this property is concerned is the transposon, Tn7, which has evolved alterative strategies to ensure/promote its propagation. In one pathway, Tn7 like many other TEs, can transpose to many different sites albeit at low frequency. Although these sites are unrelated in their DNA sequence, there is a marked preference for conjugal plasmids of broad host range. In other pathways, transposition takes place in a single chromosomal site, called *attTn7* or its attachment site, found in diverse bacteria. In order to operate the two selection choices, Tn7 codes for two site selector proteins, TnsD and TnsE, as described earlier. TnsE directs the Tns ABC transposition machinery to various sites in conjugal plasmids. This occurs in a single orientation targeting DNA undergoing lagging strand synthesis in recipient cells. TnsD, on the other hand, is a sequence-specific DNA binding protein that recognizes and binds to *attTn7*. You may recall that this sequence is located downstream of the *glmS* coding region. TnsD binding induces an assymetrical distortion at the 5′ end of binding at *attTn*, that helps in recruiting the TnsC at this site. TnsC, therefore, forms a platform that could receive the transposase, TnsAB.

7.19 Conjugative Transposons

Several bacteria possess elements that may integrate and excise, like a prophage, as also the elements that are transferred by conjugation, such as plasmids. Subsequently, however, a number of such elements were identified which combined these two properties, such that they are excised by site-specific recombination and transfer the resulting circular form by conjugation. The latter integrates by site-specific recombination between a specific site it carries and a site in the genome of their host. These elements form a heterogeneous group and are referred to as conjugative Tn, conjugative 'plasmids', and genomic islands, etc., and all of them have been classified as integrative and conjugative elements (ICE).

These elements are of chromosomal origin and code for site-specific recombinases that promote their excision and integration. Different types of site-specific recombinases are known, such as those catalyzing integration and excision of circular molecules (integrases), those resolving co-integrates (resolvases), and those bringing about inversion of specific DNA segments (invertases). They have been broadly classified into two unrelated families: serine recombinase and tyrosine recombinase family.

The best-studied chromosomal conjugative elements, though many others are known now, are the conjugative transposon Tn916 from *Enterococcus faecalis* (Fig. 7.12) and pSAM2 from *Streptomyces ambofaciens*. The latter was first described as integrative plasmid.

Tn916, like all other conjugative elements from Gram$^+$ bacteria, does not code for pili instead encodes an integrase belonging to the tyrosine family. Tn916 conjugative transfer involves its excision catalyzed by the integrase and excisionase that creates a covalently closed circular molecule. The latter carries an attachment site *attTn* resulting from recombination of two attachment sites, *attL* and *attR* flanking Tn916. This is followed by its conjugative transfer to a recipient cell that may be similar to that of conjugative plasmids (Fig. 7.13). Tn916 also codes for all the proteins needed to form mating pore and mediate its own transfer. This involves a nick at the *oriT* so as to initiate the transfer of a single strand of the circular intermediate. After transfer and replication, both donor and the recipient will contain a copy of the double-stranded circular molecule. In the recipient cell, the integration of the circular form takes place at a number of sites in *E. faecalis* and in numerous other bacteria preferentially in the A-T-rich region. This reaction is catalyzed by a tyrosine recombinase. Site-specific excision and low specificity level of integration in the same cell can also take place resulting in intracellular transposition. The transposition mechanism is very different from that of IS and type I and type II Tn which involves a typical DDE transposase.

Many other conjugative Tns may show higher (Tn5276 from Gram$^+$ *Lactococcus lactis*) or lower specificity (cTnDOT or Tc^r ERL from *Bacteroides*) of integration. Conjugative Tn, CTnscr94 from *Salmonella enterica*, Tn5252 from *Streptococcus pneumoniae*, *bph-sal* from *Pseudomonas putida*, *clc* from *Pseudomonas* sp, symbiosis islands from *Mesorhizobium loti*, and SXT from *Vibrio cholerae* are site-specific integrative and conjugative elements though many of them were initially not classified as conjugative Tn. The mechanism of conjugation is not fully known but supportive evidence could be derived for many. For example, SXT could code for pili and the origin of conjugational transfer could be identified in Tc^{r}Em^r DOT from *Bacteroides*, Tn5252, and SXT elements.

A large variety of ICEs are integrated in a site-specific manner, though in some cases this integration may be host-dependent. This would mean that in one host the integration may exhibit high specificity, whereas in another host the same element may show low specificity. Very often the integration site for many such elements is the 3′ end of a tRNA-encoding gene. Again,

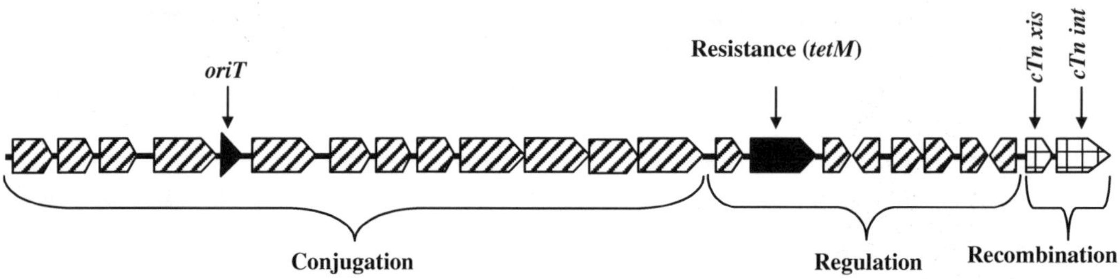

Fig. 7.12 Schematic representation of structural organization of conjugative transposon, Tn916. Orfs identified constitute three regions that have been implicated in conjugation, along with *oriT*, regulation, and recombination (those coding for excisionase, and integrase). Tn916 also carries tetracycline resistance gene

Fig. 7.13 Proposed steps in the excision and integration of a conjugative transposon, e.g., Tn916. The integrated element is excised by staggered cuts that take place six nucleotides from the ends of the element producing single-stranded overhangs or coupling sequences. The coupling sequences are joined covalently to produce a circular transposition intermediate. The bulged area indicates the region where base pairing does not take place. The circular form and the new target site are both opened up by staggered cuts and ligation produces two non-base paired regions. The latter are resolved either by mismatch repair or replication through the region

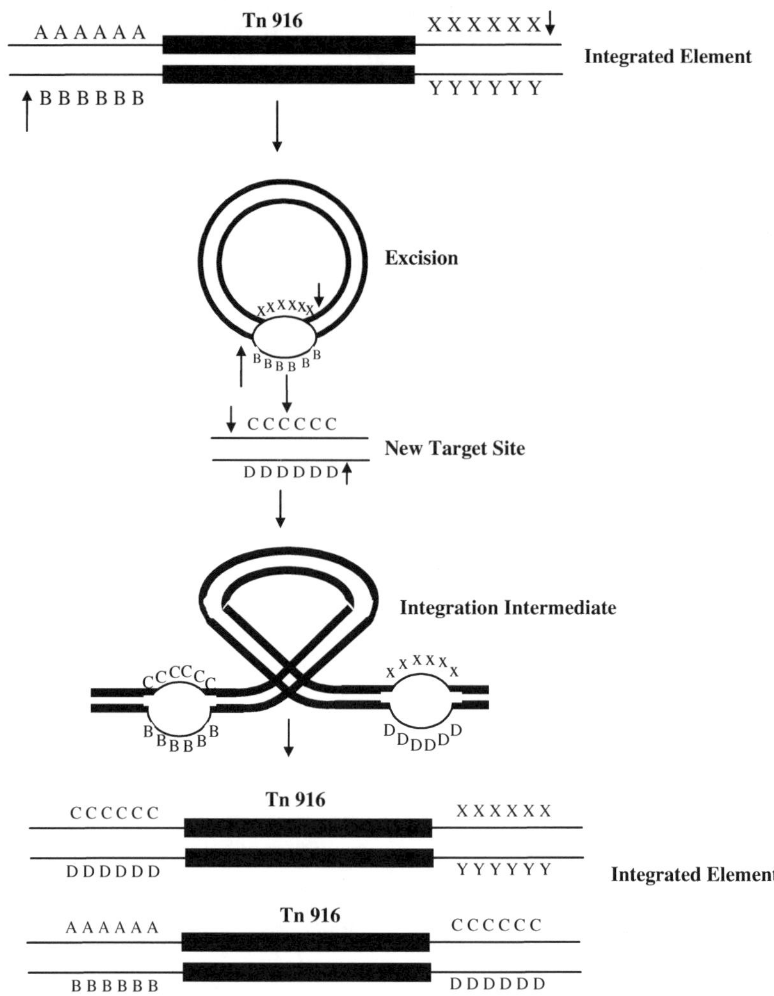

an element like pSE101 from *Saccharopolyspora erythraea* integrates at 3′ end of $tRNA^{Thr}$, but in *Streptomyces lividans* it uses several other sites. As these elements do have various integration sites, they are also able to transpose within the genome of their host (intracellular). Almost all site-specific integrative and conjugative elements/conjugative Tn encode a tyrosine recombinase, a feature that they share with integrative prophage. The one exception to this is the Tn5393 that codes for serine recombinase. On the basis of sequence analysis, however, some elements with serine recombinase have also been identified from *Clostridium difficile, E. faecalis,* and *Streptococcus*.

Although when first identified the ICEs/conjugative Tns were thought to be of limited occurrence, they now appear to be widely distributed. This was based on the fact that many of these elements were first described as conjugative/integrative plasmids. The confusion between these elements and plasmids prevailed for a long time because these elements could spread by conjugation (R391 from *Providencia rettgeri* and R997 from *Proteus mirabilis*) or could mobilize the coresident plasmids by co-integration (pRS01 from *Lactococcus lactis* ML3) or were isolated as circular form (R391, R997, and pRS01), or a *oriT* was identified on them (pRS01).

The high G+C Gram-positive bacterium *Streptomyces* and its relatives *Saccharopolyspora* and *Amycolatopsis* possess several ICEs. The best studied of these elements is pSAM2 from *Streptomyces ambofaciens*. This element integrates into a unique site identified as 3′ end of $tRNA^{Pro}$ gene. Its circular form could only be detected in the presence of a recipient mycelium. The element carries *repSA* (rolling circle replicase), *xis* (excisionase) and *int* (integrase) genes organized in a single operon and activated by *pra* gene product. Interestingly, *pra* transcription is repressed in the absence of the recipient mycelium ensuring that the wild-type pSAM2 remains integrated. A *pra* promoter mutation on pSAM2 derivative defective in integrative functions (*attP* deletion or *int* gene disruption) results in free replicons.

A physical contact between the donor and the recipient mycelia leads to activation of *repSA-xis-int* and consequent excision and rolling circle replication of pSAM2. Its inter-mycelial transfer is mediated by *traSA* and the transferred copy is replicated in the recipient. The intra-mycelial transfer within the recipient is mediated by four genes *spdABCD*. Although several similarities can be drawn between pSAM2 and plasmids in terms of transfer proteins and physical appearance, pSAM2, unlike a plasmid, is found integrated in the donor and the recipient mycelia. Although pSAM2 is replicated the replication of circular form is not involved in the maintenance (as would happen with the plasmids), but is necessary for its transfer.

7.20 Integrons

One of the most striking examples of bacterial evolution has been the emergence and spread of antibiotic resistance among different bacteria, especially the pathogens. While the first event was manifested by the selection provided through the use of antibiotics, the other was traced to the presence of extra-chromosomal elements, such as plasmids and TEs. Most importantly, within the few years of the usage of antibiotics, multiple drug resistance emerged amongst several pathogens. This attracted the attention of microbiologists/microbial geneticists/clinical microbiologists alike, because it threatened the infectious disease therapy in a big way. While single-drug-resistance phenotype is understandable, multiple drug resistance was not entirely anticipated because it requires co-appearance of multiple mutations that would simply be beyond the evolutionary potential of a given bacterial population.

When many of these multi-drug resistant strains were isolated and analyzed it became clear that this could not be due to mutation alone. One important association was revealed between transmissible plasmids and/or Tn with such a resistance. A new dimension was added when gene-capture platforms or integrons and their ability to acquire genes (conferring resistance or any other phenotype of selective advantage) was reported. The involvement of an integron in a transposon, such as Tn21, and its ability to express antibiotic resistance phenotype provided the lead toward the development of multidrug resistance. Analysis of multidrug resistance in *Shigella dysentriae* made it clear that bacteria were thoroughly prepared to face

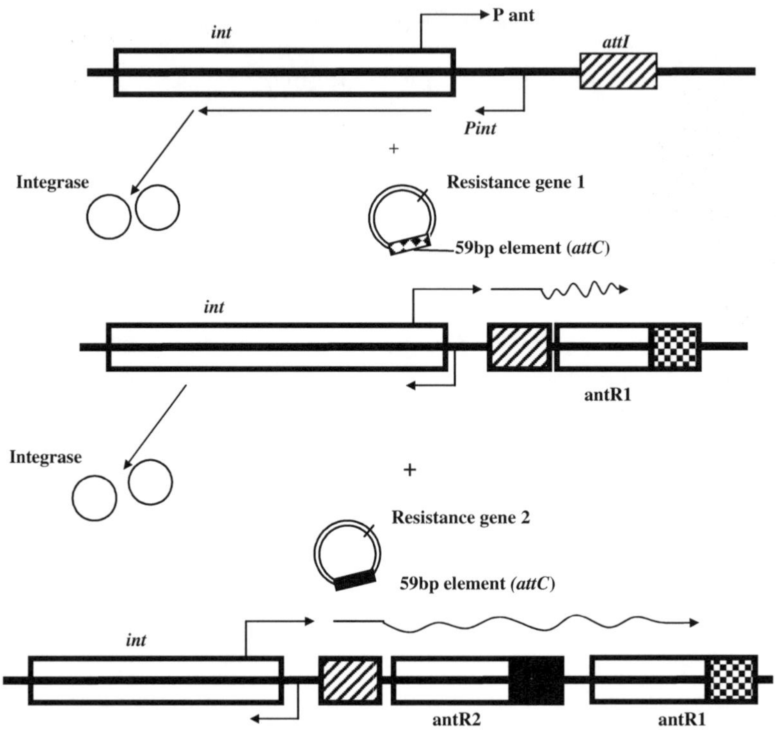

Fig. 7.14 General plan of resistance gene capture by integrons. The basic unit of an integron consists of a gene encoding a site-specific integrase (Int) and a site-specific integration site (*attI*). A resistance gene (ant R1) in the circular cassette form containing the 3′-terminal 59-bp element or *attC* can be integrated at *attI* with the help of integrase. The *attC* site may be highly variable in length and sequence. Integration site so generated provides acceptance site for another resistance gene (antR2). Such integrations can continue leading to multiple antibiotic resistance. The resistance genes are all integrated in the same orientation and are transcribed by the strong promoter, Pant located in the integron

the onslaught of different antibiotics and quickly responded by evolving appropriate genetic tools such as integrons.

Integrons are essentially gene capture units that provide the assembly of exogenous ORFs by site-specific recombination and convert them into functional genes by ensuring their appropriate expression. Thus, all known integrons consist of three essential elements: a gene *intI* encoding an integrase belonging to tyrosine recombinase family; a primary recombination site *(attI)*; and an outwardly oriented promoter (Pc) that directs the expression of the captured gene/(s) (Fig. 7.14). Integrase can insert discrete units of circularized DNA known as gene cassettes in a Rec-A independent manner. The integration occurs at *attI,* downstream of the resident promoter Pc, such that expression of the genes in the cassette could be effected. The gene cassettes share many characteristics that comprises a single gene and an imperfect inverted repeat at the 3′ end of the *attC* site also referred to as 59-base element (59 be). The *attC* sites, however, are a diverse family of nucleotide sequences that function as recognition sequences for the site-specific intergrase. They may range from 57 bp to 141 bp in length and their nucleotide sequence similarities are essentially restricted to the boundaries which contain conserved sequences, known as the inverse core site (ICS) or R″ sequence (RYYYAAC) and the core site (CS) or R′ sequence (GTTTRRY), where R is a purine and Y is a pyrimidine. The point of recombination lies between G and the T bases.

All integrons are broadly classified into two distinct classes:

(1) The mobile integrons, though they themselves are not mobile but are linked to a mobile DNA element, and
(2) The superintegrons, which carry a large number of gene cassettes associated with the integron. The *attC* site of these cassettes carries a high degree of identity and the structure is neither mobile nor associated with a mobile element.

Mobile integrons have been classified into five classes based on the sequence of their integrases (40–58 % identity). All are linked to mobile elements such as ISs, Tn, and conjugative plasmids. These can serve the purpose of intra- and inter-species transmission of genetic material such as the spread of antibiotic resistance genes. Class 1 integrons are associated with functional or non-functional Tn derived from Tn402 that can be embedded in larger Tn, such as Tn21. Class 2 integrons, on the other hand, are exclusively associated with Tn7 derivatives, as described earlier, and Class 3 integrons are located in a transposon inserted in a plasmid. Class 4 and class 5 integrons are involved in the development of trimethoprim resistance in *Vibrio* sps, with class 4 forming a component of SXT element in *V. cholerae* and class 5 a part of compound transposon carried on a plasmid in *V. salmonicida*.

Class I integrons are more common among clinical isolates, with over 80 different antibiotic-resistance gene cassettes. These elements between them are able to confer resistance to all β-lactams, all aminoglycosides, chloramphenicol, trimethoprim, streptothricin, rifampin, erythromycin, fosfomycin, lincomycin, and quaternary ammonium compounds-based antiseptics. Many class I integrons are associated with ORF 513 coding for a Tpase belonging to IS91 family. This large element is also referred to as complex *sul1*-type integron.

That the type I integrons provide a gene-capture platform can be ascertained by the differences in the codon usage among gene cassettes even in the same mobile integron. By acquiring genes of diverse origin, integrons have helped the bacteria build-up a stockpile of different genes that may provide an adaptive edge to the host under certain sets of conditions. Mobile integrons have been described from both Gram-negative and Gram-positive bacteria, suggesting thereby that they have served a common purpose in bacterial evolution.

In the later years, a cluster of repeated DNA sequences with an organization similar to mobile integrons was described from chromosome 2 of *Vibrio cholerae* called *Vibrio cholerae repeats* (VCR). These are large elements carrying >20 gene cassettes with a high degree of identity (>80 %) between *attC* sites of their endogenous cassettes. The integron codes for a specific integrase, Vch Int1A that is related to the integrase of mobile integrons. Yet, this can be differentiated from the latter in having high homology between the *attC* sites and it does not seem to be mobile. The latter emanates from the fact that it is located on the chromosome and is not associated with a mobile element. These elements were called superintegrons.

Although first described from *Vibrio cholerae*, superintegrons have been identified from many other members of Vibrionaceae, Xanthomonads, and Pseudomonads, and other members of γ-proteobacteria. Subsequently, on the basis of homology, ORFs coding for integron-associated integrases have also been detected in α- and β-proteobacteria and many others. While many show characteristics common to other superintegrons, some have departed from the conventional types. Analysis of the biological diversity of the metabolic abilities associated with superintegrons (including antibiotic resistance and virulence) suggests that these gene-capture platforms have helped the bacterial host especially in the context of a survival-enhancing stress response.

Comparative analyses of integron-associated integrases, both characterized and hypothetical, show that they form a specific clade within the tyrosine recombinase family. Species-specific clustering of the integrase genes reflects the line of descent among the bacterial species carrying them, and puts the establishment of superintegrons much before the speciation in the respective genera. Thus, integrons are ancient structures that profoundly impacted the bacterial genome evolution. The ubiquitous nature of the integrons in the genomes of the environmental bacterial species derived from markedly different sources suggests that they have served an important role in the emergence of novel bacterial species. As gene-capture platforms, they helped in the acquisition of genes having adaptive advantage and enabled the host to respond to a particular environmental condition quickly. While transpositions could provide one pathway for gene clustering, horizontal gene transfer has also played an important role in gene cassette trafficking. Biological diversity of the captured genes has the potential to discover novel proteins. The proposed antiquity of the superintegrons has led to the suggestion that mobile integrons are derived from them. This would have involved the entrapment of *intI* genes and their cognate *attI* sites by mobile elements reminiscent of the evolution of compound Tn. The possibility has also been forwarded that the assembly of complex integrons (carrying more than two resistance cassettes) may also occur by recombination between two mobile integrons rather than by successive recruitment events. Several environmental niches other than the obvious clinical settings have been identified, in which such an exchange of gene cassettes between mobile integrons through horizontal gene transfer can be facilitated.

7.21 Transposable Elements as Genetic Tools

TEs are remarkably diverse molecular tools that can be applied to single-gene as well as whole-genome studies. Endowed with inherent mobility they are apt at reshuffling their genomic location without any need for sequence homology. This one property enables them to modulate the gene expression thus contributing immensely to genome plasticity and evolution. From both the academic and applied points of view, TE most favored as genetic tools are those that insert randomly or near-randomly, or can be manipulated to behave in this way. Also, those carrying a selectable marker (e.g., antibiotic resistance) were preferred over others as the mobility of transposon could be monitored easily. While their role in introducing random mutations and chromosomal aberrations is still being sought after by geneticists, they have also been put to several other usage. We shall discuss some of these applications in the following discussion.

7.21.1 In Vitro Transposition

Detailed mechanistic analysis of transposition demanded the development of in vitro transposition reaction which is now available for various elements such as Tn3, Tn5, Tn7, Tn10, Tn552, IS911, and bacteriophage Mu to name a few. This has been extended to eukaryotic Tn also such as Ty1 of yeast and mariner of insects. In vitro transposition reaction constitutes the transposon's terminal inverted repeats, purified transposase, the DNA target substrate, in a simple reaction buffer. When standardized this reaction proceeds with high efficiency producing thousands of transformants per microgram of substrate plasmid DNA. While this process revealed some very useful information on transposition process per se, it also provided a way to generate genomewide insertion mutations in bacteria and yeast. Essentially, this method involves the purified genomic DNA of the target organism, subjected to in vitro transposition, and that is then transformed into the host with the selection for the transposon-associated marker. Homologous recombination within the cell results in replacement of the wild-type gene with the mutated allele.

7.21.2 Gene Fusions

The random insertional property of many Tns has also been exploited in generating transcriptional and translational fusions. Such fusions are directed toward a target gene for which the Tn is modified to carry a reporter gene. Two very popular reporter systems have been *lacZ* and *phoA* because of the ease with which their respective products, β-galactosidase and alkaline phosphatase, can be chromogenically quantitated. Subsequently, other reporter genes such as *gfp* and *lux* have also been introduced. In these fusion systems the Tpase is generally supplied in *trans*. This not only minimizes the size of the vector but also allows the regulated expression of the Tpase gene and reduces the frequency of secondary transposition events.

Translational fusions such as in Tn5-*phoA* is designed by truncating the 5′signal peptide sequence of *phoA* and fusing it in-frame to one end of the Tn. When such a Tn gets inserted in-frame in a target gene it will produce a fused protein consisting of a portion of the N-terminus of the target protein, a short peptide derived from the Tn end, and the bulk of PhoA. PhoA is active mainly in the periplasm. Thus, if the N-terminus region of the target protein includes signal for periplasmic location, the fusion protein will be transported out of the membrane and the alkaline phosphatase will be activated. The enzyme activity can be monitored with a chromogenic substrate. Such a modified transposon has helped in the genetic dissection of the gene (s) whose products are either exported or involve cell membrane/envelope architecture.

The Tn, such as Tn5, Tn10, designed to carry *lacZ* gene have been used to produce transcriptional fusions. Any in-frame insertion of this Tn next to a promoter can be ascertained by *lacZ* expression (β-galactosidase plate\enzyme assay). These studies have helped in determining the strength of the promoter under different physiological, environmental, and genetic conditions.

Reporter fusion technology along with other Tn-based methods will remain an invaluable tool in providing important information both at single-gene and whole genome levels.

7.21.3 Signature-Tagged Mutagenesis

Genetic dissection of many complex functions such as microbial virulence in animals require negative selection of attenuated mutants. Tn-based signature-tagged mutagenesis (STM) has turned out to be a useful tool for large-scale mutagenesis, recovery of a range of avirulent mutants, and their in vivo screening. In STM, a pool of synthetic oligonucleotides with random internal sequences (~40 bp) but with invariable 3′ and 5′ ends are cloned in a transposon delivery vector. The pool of oligonucleotides comprises <100 unique tags that not only do not cross-hybridize in subsequent step but also PCR amplify reproducibly. The tagged Tn pool is used to randomly mutagenize the genome of interest and Tn mutants are arrayed on plates for later screening and recovery. Pools of mutants are used to infect an animal or cell culture model. After the infection is allowed to proceed for required period of time, bacteria are recovered from selected tissues and DNA is prepared from both the recovered bacteria (output pool) and the inoculum (input pool). The tags present in the two genomic preparations are PCR amplified using the invariable ends of the tag as priming sites, radioactivity labeled, and hybridized against the colony arrays or DNA dot blots thereof. Mutants that fail to hybridize with the output pool probe are identified as the likely candidates, unable to establish themselves in the animal and thus must be impaired in virulence. These mutants are further tested to prove their role in pathogenesis.

One of the most commonly used STM employs a mini-Tn5 Tn and has proved its utility in a number of pathogenic proteobacteria. STM has turned out to be so useful that it has been further employed in Gram-positive bacteria by replacing Tn5 with Tn917, Tn1545, and IS1096 as the former transposes poorly in this group. This technique has also been extended to the development of novel antimicrobial agents and vaccines.

7.21.4 Genetic Footprinting

Like STM, genetic footprinting has been used to study the essentiality of certain genes under experimental or test conditions. Often these genes belong to the category of those involved in growth and viability, where subtle competitive disadvantages cannot be manifested through traditional mutagenesis. The first step involves the generation of a large pool of insertion mutants and subjecting the mutagenized pool to grow under the test conditions for a specific number of generations. Genomic DNA is retrieved from the pool before and after growth and is subjected to PCR using a pair of primers, one that anneals to the transposon ends and the other that primes from within or nearby the gene of interest. The depletion of PCR product following growth under the experimental conditions is indicative of a decline in the number of cells in that population harboring the transposon insertions in the gene of interest.

The genetic footprinting strategy has proved exceptionally useful for analyzing single as well as a spectrum of genes both in bacteria and yeast.

7.21.5 Gene Sequencing

The inherent ability of Tn to insert into different sites enables them to provide the priming sites for nucleotide sequence determination of uncharacterized genomic regions. Sequencing primers that anneal near the ends of a Tn from a collection of Tn insertions into a cloned DNA fragment can generate a set of overlapping sequence contigs. These can subsequently be assembled into the whole sequence of the DNA fragment. A number of Tn such as Tn3, Tn5, and Mu have been developed as sequencing tools.

In another variation, sequencing with Tn-specific primers has been used to localize the insertion sites directly in genomic DNA of yeast and bacteria.

7.21.6 Transposons and Microarray Technology

Microarray technology has ushered in a study of genome-wide gene expression in a variety of prokaryotic and eukaryotic systems. Combining with the power of random insertion mutagenesis it can lead to the identification of genes that contribute, say for example, toward the fitness of cells growing under certain conditions or genes required for biosynthesis of essential metabolites under such a condition. These can be further resolved through microarray analysis.

7.21.7 Scanning Linker Mutagenesis

Many a time, the insertion of short peptides in the construction and analysis of proteins' structure–function relationship provides a valuable tool. This strategy is called scanning linker mutagenesis (SLM) that can be achieved by insertion and imprecise excision of the transposon from a corresponding cloned gene. The excision is often achieved by digesting the Tn-containing (inserted in the gene of interest) plasmid by a restriction enzyme that cleaves near to both the ends of the element. Rejoining of the restricted fragment bearing the target gene reconstitutes the plasmid but with an in-frame insertion in the gene.

7.22 Transposable Elements and their Impact on the Host

Looking at the multitude of the applications, various Tns have been modified to generate derivates, suited for a particular application. While we have discussed various such modifications above, one element that is of widespread use is the transposable phage Mu. Mu genome has been manipulated by curtailing its phage genes, but retaining the transposition required elements (Mu ends, MuA, sometimes MuB, and *pac* site) into mini-Mus. They have been further modified by the introduction of a selectable marker, such as drug resistance and/or reporter gene(s). Some Mu-based vectors contain a thermo-sensitive allele of repressor gene, *c*. Mu derivatives have also been designated as MudI and MudII, respectively. Mu elements and mini-Mu plasmids have been extensively employed as tools of in vivo and in vitro genetic engineering.

Conjugative Tns are the important determinants of antibiotic resistance, mainly in Gram-positive bacteria. They are highly useful as they provide a means for conjugation between bacteria belonging to different species and genera with remarkable promiscuity. The mechanism of conjugative transposition differs from other Tns in several details. The site-specific recombinase, Int, belonging to integrase family and the excision protein, Xis, bear striking resemblance with phage λ system but recombining sites are different in the conjugative Tn, as described earlier.

A large number of TEs are normal constituents of bacterial genomes and plasmids, suggesting that these elements have played a significant role in bacterial evolution. Another striking property of these elements is the generation of gross chromosomal rearrangements that has added onto the genetic variability in their hosts.

From the time that mobile genetic elements were described, TEs have been found to be associated with a variety of chromosomal rearrangements, such as deletions, duplications, inversions, the creation of acentric and dicentric chromosomes, translocations, and recombination of host genomes. All these have contributed to the genome plasticity and have significant implications in biological evolution.

Two possible mechanisms by which TE-mediated rearrangements can occur are:

(i) Indirectly by homologous recombination, and
(ii) Directly by an alternative transposition process.

As discussed earlier, TEs can generate multiple copies of similar if not identical sequences that can provide potential targets for homologous recombination. In addition, faulty repair of double-stranded breaks introduced during TE excision can lead to formation of footprints. Homologous recombination can result in duplications, deletions, and inversions depending upon the relative orientation of these elements.

Another mechanism that leads to chromosomal rearrangements is through alternative transposition. Here, the complementary ends from separate TEs synapse, generating a hybrid element referred to as bimolecular synapsis. Excision of such a hybrid element results in two double-strand breaks which will be repaired. The excised hybrid element can also, however, reinsert in the genome. The chemical steps in alternative transposition are identical to those of a normal transposition and a Tpase is required. However, unlike the normal event, one end of each hybrid element remains covalently bound to a chromosomal fragment. Most of the reported chromosomal rearrangements are deletions, duplications, and inversions, and in eukaryotes, dicentric and acentric chromosomes can also be formed.

Further Readings

Faelen M, Toussaint A (1976) Bacteriophage Mu-1: a tool to transpose and to localize bacterial genes. J Mol Biol 104(3):525–539

Goldhaber-Gordon H, Early MH, Gray MK, Baker TA (2002) Sequence and positional requirement for DNA sites in a Mu transpososome. J Biol Chem 227: 7703–7712

Gray YHM (2000) It takes two transposons to tango. Transposable element-mediated chromosomal rearrangements. Trends Genet 16(10):461–468

Haniford DB (2006) Transpososome dynamics and regulation in Tn10 transposition. Crit Rev Biochem Mol Biol 41:407–424

Hayes F (2003) Transposon-based strategies for microbial functional genomics and proteomics. Annu Rev Genet 37:3–29

Mahillon J, Chandler M (1998) Insertion sequences. Microbiol Mol Biol Rev 62(3):725–774

Mazel D (2006) Integrons: agents of bacterial evolution. Nat Rev Microbiol 4:608–620

Nagy Z, Chandler M (2004) Regulation of transposition in bacteria. Res Microbiol 155:387–398

Peters JE, Craig NL (2001) Tn7: smarter than we thought. Nat Rev Mol Cell Biol 2:806–814

Reznikoff WS (2003) Tn5 as a model for understanding DNA transposition. Mol Microbiol 47(5):1199–1206

Rice LB (1998) Tn916 family conjugative transposons and dissemination of antimicrobial resistance determinants. Antimicrob Agents Chemother 42(8): 1871–1877

Rowe-Magnus DA, Mazel D (2001) Integrons: natural tools for bacterial genome evolution. Curr Opin Microbiol 4:565–569

Salyers AA, Shoemak NB, Stevans M, Lhing-Yew L (1995) Conjugative transposons: an unusual and diverse set of integrated gene transfer elements. Microbiol Rev 59(4):579–590

Shapiro JA (2009) Letting *Escherichia coli* teach me genome engineering. Genetics 183(4):1205–1214

Starlinger P, Saedler H (1976) IS elements in microorganisms. Curr Topics Microbiol Immunol 75: 111–152

Vizvaryokova M, Valkova D (2004) Transposons—the useful genetic tools. Biologia Bratislava 59(3): 309–318

Additional Readings

Some Books in the Subject Area

Stanely RM, John EC, David FJ, Bartlett (1994).Microbial genetics, 2nd edn

Nancy T, Janine T (2003) Fundamental bacterial genetics, Blackwell

Ulder NS, Ronald EY (eds) (2002) Modern microbial genetics, 2nd edn. Wiley-Liss

Larry S, Wendy C (2012) Molecular genetics of bacteria, 3rd edn. 2007/4th edn. with two additional authors: Peters JE, Henkin TM ASM Press

Jeremy MD, Simon PP (2010) Molecular genetics of bacteria, 5th edn. John Wiley

John S, David L, Galizzi A (eds) (1985) Genetics of bacteria, Academic Press

Edward AB (2006) Bacterial and bacteriophage genetic, 5th edn. Springer Science + Business Media Inc

William H (1965) The genetics of bacteria and their viruses, John Wiley

Neidhardt FC et al.(eds) (1996) *Escherichia coli* and *Salmonella typhimurium*, vols 1 and 2. Cellular and molecular biology, 2nd edn. ASM Press

Krebs JE, Goldstein E, Kalpatrick ST (2011) Lewin' genes X, Jones & Bartlett Publication

David PC, Pazdernik NJ (2012) Molecular biology: understanding the genetic revolution, 2nd edn. Elsevier

Griffith AJF, et al (2008) Introduction to genetic analysis, 9th edn. WH Freeman and Company

Krebs JE, Goldstein ES, Kilpatrick ST (2010) Lewin's essential genes, 2nd edn. Jones and Bartlett Publication

Klug WS, Cummings MR (2003) Concepts of genetics, 7th edn. Pearson Education Inc

Sambrook J, Russel DW (2001) Molecular cloning: a laboratory manual, 3rd edn. Cold Spring Harbor, New York

Index

R

Printed by Printforce, the Netherlands